PACIFIST TO GLIDER PILOT

Revised edition published 2001

First published in 2000 by
Woodfield Publishing, Bognor Regis,
West Sussex PO21 5EL England

ISBN 0 873203 53 5

Pacifist to Glider Pilot

Alec Waldron

Woodfield Publishing
~ WEST SUSSEX · ENGLAND ~

The steamer aboard which my grandparents sailed to India in 1890 (see page 14).

Contents

The Author in the uniform of the Glider Pilot Regiment, 1945.

About the Author

Alec (christened Albert) Waldron was born in Copnor, Portsmouth on 6th October 1920. He attended Portsmouth Northern Secondary School from 1932 to 1937, when he left home to join the Admiralty, London as an accounts clerk.

Promoted to Executive rank in 1940 and volunteered for service in the Royal Corps of Signals in November 1940 and transferred to the Glider Pilot Regiment in January 1942 for training as a Glider Pilot. In July of that year was a member of the 1st Course at Brize Norton to convert onto the Horsa glider.

Following the successful completion of this course was promoted to Staff Sergeant and was posted to HQ, 1st Battalion Glider Pilot Regiment, to set-up an Intelligence Section, whilst retaining his duties as a pilot. In this capacity he was involved in the planning of the Sicilian Airborne Operation of July 1943, in which he also participated as a pilot. He forced landed in the sea, was wounded, evacuated to Malta for treatment and finally returned to his Battalion. Subsequently he received a Certificate of Commendation from General Montgomery. In the following August he took part in the landing at Taranto, Italy by the 1st Airborne Division – a seaborne landing!

He returned to the UK in January 1944 and was retained as Staff-Sergeant, Intelligence though by this time the 1st Battalion had been renamed Number 2 Wing (!). He was involved in operational planning up to and including the Airborne attack on Arnhem in which he participated as a pilot. During the battle that followed the succesful landings, his regiment lost every HQ officer and the RSM, killed or wounded, and thus he took over duties as the senior survivor at HQ. He was wounded but returned safely the the UK and was awarded a Mention in Despatches for 'services rendered'.

In February 1945 he was given an 'Immediate' Commission on the instructions of the Under Secretary of State for War, promoted Captain on 1st June 1945 and in January 1946 was appointed Staff Captain (GSO III) with special responsibilities to organise and run all the Glider maintenance and flying activities at RAF, Syerston then designated 1333, Transport Support Training Unit. He was demobbed in October 1946.

After demob he opted not to return to the Civil Service but enrolled in a Government sponsored Industrial Training Scheme for 2½ years at Parsons Chain Co, Stourport-on-Severn and was subsequently appointed Production Manager of that Company. He also successfully contested a local Urban District Council election as a British Legion candidate pledged to secure a fair deal for ex-servicemen seeking Council houses. He became Chairman of Housing and initiated and organised the first ever housing estate to be built on 'stabilised soil' rather than orthodox concrete as a means of cost saving. The Estate still stands.

After 7 years on the Council he resigned and took up a post with Stewart and Lloyds near Liverpool. Two years later joined Addis Ltd, a family Company of nearly 200 years standing, producing brushware and plastic goods, first as Production Manager and later, in 1968 as Works Director with responsibility for five factories. Later he combined these duties with those of Joint Managing Director. He retired in October 1980.

FOREWORD

A Fickle Handmaiden

The 'journey' portrayed in the following pages, from 1920 to 1946 and stretching over a third of my seventy-eight years, was in turn pleasurable, painful : happy, sad : but certainly tortuous. It lacked any significant career ambitions save to achieve job security and a reasonable income, prime objectives set by my father who was determined that we children would not have to endure his problems of low pay and uncertainty. It also lost an all-embracing focal point when I discarded the 'anchorage' provided by the Exclusive Plymouth Brethren. Indeed, my reaction to this traumatic experience brought about by my decision to volunteer for Army Service - initially in the Royal Corps of Signals - in World War 2 was, first one of relief, soon to be followed by a search for an alternative belief outside the tenets of Fundamentalist Christianity. Undoubtedly the shock transition from Civilian to Army life, with the consequent need to wrestle with the loss of a personal identity, fatigues, drill, guard duties and the like, pushed my theological and philosophical concerns to the back of my mind and thus, in an odd kind of way, gave me a much needed emotional breathing space. Within days of my enlistment an even greater and more intense personal shock was to hit me. My father, a pacifist by religious conviction, was killed by a bomb whilst cycling home to be with my mother during a hit-and-run raid on Portsmouth. This was December 1940 and, unknown to me, the beginning of six years of a formative and at times fascinating Army life which was to shape my future beyond 1946.

To glorify my attempts by describing this 'journey' as an autobiography would be arrogant in the extreme. I am no diarist so must rely on that most fickle of handmaidens - memory. Inevitably inaccuracies, embellishments and memory lapses will permeate my story but hopefully the resultant freedom from a strict historical discipline will enable me to be deliberately selective both in content and emphasis. Likewise, freed from the constraints of writing a day by day story in a chronological sequence, I have at times set out to paint a series of vignettes - windows - on different aspects of my life, frequently overlapping in time but each a significant ingredient in my personal journey from 'toddler' days to demob as a Captain in the Glider Pilot Regiment.

In no sense however should the subsequent pages be read as setting out a deliberately planned path of renunciation, liberation and career development but rather as a seemingly random metamorphosis from pacifist to pilot, which though ending in 1946 provided the 'launching pad' for the ensuing thirty-four years both in Industry and somewhat briefly, in politics.

CHAPTER 1

The Wind is in the East

The wind, blowing in from Langstone Harbour over Salterns and the 'Corporation Fields', eddied and swirled as it rebounded off the opposite side of our terraced street carrying dust from our dried-out gravel road into the minuscule porch of our end-of-the-terrace house. My industrious and house-proud mother waged an unceasing and fruitless 'war' against this invasion of our cleanliness, complaining bitterly to my father that the wind was once again in the East and therefore high time that the Portsmouth Corporation brine laden water-cart paid Martin Road, Copnor, a visit to sprinkle the gravel and thus lay the ever invasive dust. This mother and father of mine – who were they and by what tortuous paths had they come together before bringing into this world three children – two sons and a daughter – of whom I was the youngest and apparently the 'apple of my mother's eye'! Not for them the romance of childhood sweethearts nor even a long-standing 'parochial' friendship as a forerunner to eventual marriage. Such could not be, as throughout their teens they were not only some 5,000 miles apart but also growing-up in totally contrasting cultures.

The answers to these intriguing questions lie in their widely differing childhoods and inevitably call for a fleeting look at their parents' – my grandparents'- early experiences and the pattern of their lives.

Of my father's parents I remember nothing. I was 2½ years old when they died – within a few days of one another – in 1923. My grandfather, born in Ballyhaunas, Co Mayo, was a Royal Marine in his younger days but by the time he raised a family of four sons and two daughters he was an impecunious handyman, decorator, labourer, living in the Stamshaw area of Portsmouth, a virtual slum, long since demolished. My grandmother was a Berkshire woman, the daughter of a 'failed' hotelier. Family discipline was strict and poverty ever-present but they were sustained morally, spiritually and at times financially, by their membership of the Exclusive Plymouth Brethren. Despite this unfavourable background, all four sons did well at Flying Bull Lane Council School. But for a speech impediment my father would have qualified as a teacher. The impediment later disappeared but too late to rescue him from life as a poorly paid clerk with the London and Southeast Coast – later the Southern Railway.

In contrast to my father's parents, I have many memories of my mother's step-father and mother who lived on until my early teens, her own father having died in 1892. In so far as their experiences gave me, at an early age, a window into a strange and exotic world, let me recount their story in some detail as its impact on a young and impressionable child may well have provided some of the ingredients which helped to shape my future. I will start their story in the Isle of Wight during the 1880's, when my grandmother, then a housemaid in a 'stately home', was being courted by two young men – one Thomas Usher, a Lance Corporal in the 1st Hants Regiment and the other, Albert Tresize a bank clerk. The soldier prevailed and on 1st December 1886 they were married. On the 2nd January 1887 a son – Vivian – was born!

My maternal grandmother.

My mother was born on the 12th July 1888 and her sister a year and a half later. In 1890, Thomas Usher, by now a Sergeant in the 1st Hants Regiment, was posted to India and the whole family set sail by way of Aden in a leaky ship – part steam, part sail – and after an eventful voyage duly arrived in India. The ship was immediately despatched to a breaker's yard! Within two years Thomas Usher died of a fever at Barrilly, Bengal on the 18th May 1892, leaving a widow with three young children aged five, three and two years of age. What a grim situation!

Meanwhile the heartbroken bank clerk, Albert Edward Trezise having 'blotted his copybook' at the Bank, enlisted in the 1st Hants Regiment and, as a Corporal, was posted to India in 1892. 'Boy' meets 'Girl' once again and on the 16th November that same year they were married at Barrilly. There were no children by this marriage. Corporal Trezise fought in the 2nd Punjab War and was duly promoted to Colour Sergeant which doubtless indicated an excellent Service Record.

My grandmother and her family were of course 'part of the Regiment' and wherever the 1st Hants were deployed they perforce followed, travelling widely from barracks to barracks and occasionally to less established 'outstations'. The children were educated as far as possible at Roman Catholic Convent Schools where my mother distinguished herself as an exemplary pupil and a skilled embroideress. She was in fact a highly intelligent woman and had she been born in today's world would doubtless have secured an Honours Degree followed by a rewarding professional career. After an absence of thirteen years, in 1903 my grandmother and her three children returned to England and the Isle of Wight to be followed by my step-grandfather a few years later. Early in 1904 my mother, then 16 years of age, was Confirmed into the Church of England – an

The author, with his paternal grandfather...

...and his maternal step-grandfather.

early indication of a deep religious commitment and faith which was to permeate the whole of her following years. The end of the tortuous paths which finally brought my mother and father together is now in sight.

My mother's family moved to Portsmouth from the Isle of Wight where my mother was employed as a seamstress in a highly respectable dress-making establishment. Here she met amongst the employees a devout and committed Christian – a member of the Exclusive Plymouth Brethren – who persuaded her to attend one of their Meetings. In the fullness of time my mother embraced the tenets of the 'Exclusives', withdrew from the Church of England, joined 'the Brethren', met my father, already a practising 'Brother' and after a lengthy and at times a somewhat fraught courtship, they were married in an Anglican Church (yes) on the Isle of Wight on the 1st August 1912.

Little remains to be written of the years preceding my birth save to comment on the impact of the 1914/18 War on my mother and father and their two young children – my elder brother Frank born in February 1915 and my sister Ruth in October of the following year. It must have been traumatic. All Exclusive Plymouth Brethren were, and are, pacifists as a matter of religious conscience. My father was no exception. His beliefs were upheld by a Tribunal and he was conscripted into the Non Combatant Corps (NCC). This stigmatized both him and my mother. The resultant social pressures and poverty forced them to abandon their home whilst at the same time all access to my mother's parents was barred. Understandably ex-Colour Sergeant Trezise would have no brush with a 'Conchie'. My parents' unshakeable and deep religious faith undoubtedly enabled them to survive these trials and tribulations with no hint of resentment or regret.

And so to the 6th October 1920 when I was born and baptised 'Albert' – a name I detest and now supplanted by 'Alec' – as an olive branch to my step-grandfather Albert Edward Trezise. Whether this was a contributory factor I know not, but suffice it to say that the family ties were re-established, and, as a consequence tales of India, Afghanistan, Sikhs and Pathans, fakirs and rope-tricks, enlivened my childhood – another Wind from the East – giving me a window on another world.

My uncle and my father in the uniform of the World War One Non-Combatant Corps (NCC)

CHAPTER 2

Apartheid

I would that this chapter need not be written. It compels me to dredge up memories and reactions from depths which I had hoped were for ever submerged. Apartheid – yes – not South Africa but the separateness and segregation which formed such a fundamental part of my early life within the confines of the Exclusive Plymouth Brethren.

The preceding chapter looked backwards through time in order to encompass the lives of my grandparents. This will reach forward well beyond the time-span of this book – 1920 to 1946 – to reflect the impact that the Brethren exerted not only on my early life, but throughout adulthood and, to a limited extent, into my later years. In fact I shall never be able to disentangle myself entirely from the effects of 'Fundamentalism' on my early and formative years; nor should, I as there is undoubtedly a positive, beneficial side to these early experiences.

What follows is a very personal account, subjective, and in no way an objective review of the 'Exclusives', nor is it intended as an authoritative history and analysis of this sect. However some background facts are essential in order to establish them in a socio-religious context as a devout and sincere Christian Community, totally committed to their own interpretation of the Bible – both Old and New Testaments – and apparently devoid of all doubts as to the rightness of their beliefs.

John Nelson Darby (a nephew of Lord Nelson?) a clergyman in the Anglican Church of Ireland, became disenchanted with the doctrinal and organisational structures of the Church, and it is generally accepted that he set up the first Brethren Meeting Room in Plymouth circa 1831, hence their subsequent description as Plymouth Brethren. Its doctrines were grounded in Fundamentalism – a belief that every word in the Bible, from cover to cover, was absolutely true and divinely inspired. It eschewed any hierarchical concepts in its organisation, all men, though not women, being equal before God. The 'Sisters' in the Brotherhood were required to be 'subject to their husbands in all things' and in later years wore 'subjection bows' in their hair as an outward acceptance of their subordinate role in 'Meeting' affairs.

John Nelson Darby – JND as he was known – was a highly educated man and a brilliant linguist, fluent in Greek, Latin, Hebrew, French and German and probably able to converse in several other languages. In his lifetime he completed a 'New Translation of the Bible' ,"'not a revision of the Authorised Version, but a translation from the best text I (JND), attain to any certain knowledge of".The New Testament was published during his lifetime and the 'Old' after his death. Taken together with voluminous and erudite footnotes, it was intended as a study Bible rather than for public use. From 1838 to 1845 JND travelled Europe- France and Switzerland in particular- where for example, in the French Massif Central he established communities of 'Darbyites' some of which exist to this very day.

On his return to Britain in 1845 he was faced with the first major doctrinal upheaval within the Brethren concerning whether or not visitors to Meeting Rooms who affirmed their Christian faith should be permitted to partake in the Breaking of Bread (Communion) or whether such 'openness' was contrary to the Word of God. JND affirmed the latter. The sect split, the first of many such schisms up to and including the present day. Thus the two mainstreams of Plymouth Brethrenism came into being – Open and Exclusive – and it was into the latter I was born. The latter restricts its contacts to those within the sect and to gain participation in the 'Lord's Supper' as the Breaking of Bread is now known, a letter of commendation from another Exclusive Meeting is essential. Not surprisingly this' Exclusivity' has become more draconian as the years roll by and, such flexibility as existed regarding social though not religious contacts in my Brotherhood days, has long since been vetoed. This was the disciplined 'Order' within which I grew, up permeating every aspect of my life until I was twenty years of age.

Some of the Plymouth Brethren.

The Meeting Room, apart from school, was the major focal point of social and religious life. Services were frequent and, for adults, attendance was 'de rigueur'. Children, in their younger years, were excused from attending evening weekday meetings presumably on the basis of 'early to bed...' and also due to the demands of homework; education was viewed as one of the great virtues. But as one grew older and particularly during school holidays 'leave of absence' from such meetings became more difficult to obtain though my own parents were far more liberal than many in their interpretation of such requirements. To list the many Services that were held throughout the week is pointless as this varied in line with the number of Meeting Rooms within the vicinity. In Portsmouth there were three such 'Rooms' eventually increased to four when my father helped to found a new Meeting at Cosham. On Sundays there was always a morning, afternoon and evening meeting. Monday evenings throughout the whole of the sect was devoted to a Meeting for Prayer – extemporised not formal. Tuesday, Wednesday and Thursday would all provide evening meetings somewhere in Portsmouth whilst Fridays were devoted to business and organisational meetings attended only by 'Brothers' – 'Sisters' excluded. Special Tea Meetings were held from time to time on Saturday afternoons and evenings or on Bank holidays. They were dreaded by us children and on Whit Monday always avoided by my father in favour of a traditional walk for all the family except my mother. Bless him for those memorable days when he would lead us quoting Gray's Elegy from beginning to end.

Underpinning the weekly meeting routines were more personal beliefs and practices and many doctrinally based disciplines. Daily Bible readings in which all the family participated normally followed our evening meal as did family prayers, led by my father. Individual prayers were said at the bedside before retiring to rest – extemporised according to the happenings of the day and the perceived needs of the morrow. Wireless, cinema, theatre and a wide range of novels, magazines and newspapers were forbidden and it was not until my brother went to College that a daily paper – the Daily Telegraph – penetrated our household. Religious tracts such as 'Words of Grace and Comfort' were read avidly by my father and mother but found little favour with us children. Politics were never debated, and my mother and father in common with all other Exclusives, never voted. Such rules were based on matters of doctrine established by reference to and interpretation of biblical texts. Despite the original principles as established by JND vis-a-vis the absence of any hierarchical structure, inevitably as occurs in every assemblage of humans, leading Brothers emerged usually known as 'Ministering Brothers' and credited with special insights into matters divine and biblical. The 'guru' of the whole movement was one James Taylor, an American, referred to as JT, succeeded in due course by his son, JT Junior. It was from this group and mainly JT himself that new interpretations of Holy Writ were handed down and adopted as canon law. My father strongly objected to the establishment of this 'bishopric' as he dubbed it and whenever JT visited England and Portsmouth in particular, father was a 'thorn in his flesh'.

The earliest doctrinal confrontation which sticks in my memory concerned the 'Eternal Son' question, a religious conundrum which has exercised the minds of theologians as far back as the Middle Ages. In essence it asks the question whether or not Jesus Christ existed backwards through endless

The author (centre) with brother Frank and sister Ruth, c.1925.

aeons of time and then came to Earth or whether he came into Being at the time of his Virgin birth ? In due course this question became the centre of a vigorous debate within the Exclusives; postures were struck, a division emerged and some of our local Brethren 'withdrew from fellowship'. Young as I was, I tried to understand the reason for this turmoil and asked my father if it really mattered either way. Judging from his non-committal reply I like to think that he had some sympathy with my childish simplicity. Similar interpretative debates must have occurred with great frequency throughout my young life but only one other comes to mind as causing me considerable embarrassment and derision at my Secondary School. The year was either 1932 or 1933. Air travel was well established; Croydon Airport was in full swing and Airspeed in conjunction with Portsmouth City Corporation were completing the new City Airport on the North Eastern boundary of Portsea Island. Trips to the Isle of Wight by air were in the offing and all of this triggered a classroom debate. Secure in my convictions, I asserted that flying was evil on the basis that Satan was the 'Prince of the power of the air...' as stated in Ephesians Chapter 2, verse 2. I need not describe the reaction from fellow pupils and master alike. This 'law' has long since been abandoned as Brethren jet hither and thither across the face of the Earth. So much for the Divine guidance dispensed by the self appointed spiritual leaders of the sect.

But matters of far more personal significance were looming as I approached the age of twelve – the age of discretion – and all that this portended as regards Eternal Salvation. The Gospel of Luke, Ch2, verses 42 to 49, constituted the basis for the grim scenario which decreed that up to this age of discretion, although 'born in sin and shapen in iniquity', dispensation from the

My mother.

My father.

Almighty was automatically forthcoming and an eternity in Heaven was assured. From 12 years of age onwards individual salvation was dependent upon personal 'Conversion' – the recognition of one's sins and the acceptance of the blood sacrifice of Jesus Christ as the only way in which to escape eternal damnation. Frantic efforts were made by my parents to secure my conversion and I was taken to many 'hot' Gospel meetings held in one or the other of the three Portsmouth Meeting Rooms. Finally I succumbed to the emotional pressures induced by lurid word pictures of hell fire evoked by the preacher's rhetoric and to my parents' great relief I was 'saved'. Of my reactions to this event I have no recall but for the next seven or eight years I was ensconced within the tenets of the Christian faith though ever increasingly at variance with the Fundamentalism of my parents. I was a born radical and sceptic and at an early age invoked the Pauline doctrine of predestination – Romans Ch 8, verses 29 and 30 – as an excuse for my waywardness. Who gave me this questioning mind, I asked my father ?

At about this time my father, on a matter of biblical principle which I will not attempt to explain, left the Exclusives and joined the Open, to the undiluted delight of us children who thoroughly enjoyed the increased freedom that this change brought in its wake. Alas, the honeymoon was all too short lived and under theological pressure from my mother we returned to the Exclusives. Many years later my father confided in my sister that this was the biggest mistake of his life. Looking back over the intervening years I can now see with absolute clarity that my father was for ever an 'Open' spirit trapped by tradition within the straight jacket of the 'Exclusives'. Tragically he paid a heavy price for this personal conflict which he had never fully resolved.

But to return to my own personal development within the 'Faith'. Following 'conversion' the natural progression, usually in one's middle to late teens, was to ask for 'Fellowship'. However, in the Brethren this was not a religious training process but a spontaneous desire for full participation in the Christian faith as defined in the doctrines of the Exclusives entailing participation in the Breaking of Bread. Admittance into Fellowship was subject to examination and interview by three or four 'elders' of the local Meeting. I progressed to this stage in 1939 after I had left my home for London but as it was essentially linked with my adolescent love life, I will deal with this later in that context. It must be said that from my teens onwards my rebellious spirit led me ever further into deceit and hypocrisy. Visits to cinemas and theatre had to be concealed and my daily life increasingly became split into two sets of incompatible activities. Such conflicts could fill a whole book but that is not my intention.

In October 1940 a defining moment in my life was reached. Contrary to one of the basic tenets held with courage and sincerity by the Exclusives – pacifism – I decided to volunteer for HM Forces, to the great distress of my parents. This inevitably led, to my great relief, to excommunication or to use Brethren parlance, I was 'withdrawn from', an action also applied to my fiancée when she married me in July 1942. She shared my relief at these proceedings.

I wish that I could draw a line under this chapter at this point but unfortunately and distressingly the commitment of our parents to their 'faith' impinged for many years on our family lives. Throughout the post-war years we maintained contact with my widowed mother and my wife's parents, all of whom were still Exclusive Plymouth Brethren. In the late 1950's we were suddenly told, much to my wife's distress, that a planned visit of ourselves to her parents, though welcome, would necessitate separate arrangements for lunch – our family in the dining room and Grandad and Grandma in the kitchen. The new doctrine handed down by their 'leader', James Taylor was and indeed is, based on Deuteronomy Chapter 14, verses 9 and 10 referring to fish with and without scales!

We decided with great reluctance that we should not expose our young children to such inexplicable bigotry, the visit was cancelled and many tears shed. Never again did my children see their much loved grandparents. Some years later my father-in-law, then dying of cancer, moved from Kent to Glasgow in order to share a large house with my wife's younger sister who was still in the Brethren. We were living in Hertford at that time and my wife was asked to visit Glasgow as her father was clearly near to death. The night train from London took her to Glasgow where her brother-in-law declined to meet her – car sharing with an Unbeliever was not permitted – so a taxi took her to the house.

As she entered, ,my wife was faced by her sister and her two little girls. No recognition was forthcoming and the children were swept through an adjoining door before any contact could be made. Worse was to follow. After spending some hours with her father it was time for lunch. Money was proffered for my wife to go out for a meal and then return. This she declined to do and her lovely and warm hearted mother 'broke the code of conduct' and provided a light meal. For this she was subsequently severely rebuked. Weeks later her father died and despite pressure from the family we declined to attend the funeral, certain in the knowledge that at such a gathering we would have been totally ostracised. The years rolled by and mother-in-law moved back to Kent and lived with her younger son, a dedicated Exclusive. On two or three occasions she took her courage in her hands and met my wife in London. Complete secrecy was essential. She fell ill and was house-bound. All contact was lost. Letters and gifts were returned without comment, visits were not permitted to the extent that we were warned that should we present ourselves at the front door it would be closed in our faces. Of her subsequent death and burial we heard nothing for several months and then only through the good nature of a third party.

Relationships with my own mother would have followed much the same pattern but for her courage and clarity of thought. She was a highly intelligent and resolute person, difficult to live with as my late sister could testify. This sister of mine, a spinster and devout Baptist, shared a bungalow with my mother at Waterlooville near Portsmouth. As the doctrine of 'non association' – described above- was more and more rigorously enforced, mother was instructed to eat separately from my sister- dining room/kitchen arrangement once again- and to ban my wife, children and me from access to her home. This she refused to do and in due course was visited by local 'leading' Brethren. Such a visit was a prelude to her being summoned to and arraigned before a specially convened District Assembly meeting, where at the centre of encircling seats, she would have been confronted with her doctrinal failings. At such a meeting one of our distant relations lost consciousness as a result of the enormous emotional pressures thus exerted. In the event my doughty mother greeted her visitors with her Bible in hand, quoted relevant scriptures and closed the door. She collapsed with pneumonia from shock, survived, was 'withdrawn from' and at her death my sister arranged a funeral service at her Baptist Church. With her passing such episodes as I have described are, mercifully, things of the past.

Even as I near the end of this sad chapter I question my motives in placing these latter events on record. Sadness and not bitterness has motivated me as I recall these divisive experiences. As I understand it, the Exclusives are going from strength to strength into yet further extremes of 'Apartheid', imploding as they go but at the same time fracturing into yet further sub-sects. Paradoxically in today's uncertain and rootless world, such fundamentalist sects are growing throughout the Christian and Islamic communities as they offer black and white simplistic solutions to ever more complex problems.

And what of my own personal beliefs which have been subjected to the ebbs and flows of wartime and post-war experiences? Agnosticism is a woefully inadequate label with which to sum up my own personal beliefs and philosophy, insights into which will have to await the conclusion of my journey from pacifist to pilot. Suffice it to say as I conclude this chapter that for me the humility of uncertainty is an immeasurably finer attribute than the certainties of dogmatic beliefs. However, beliefs outside our own narrow self-interests are essential for the survival of the human race.

CHAPTER 3

Brass Button Alley

The fascinating world of my forbears, gleaned from old photographs, documents, scraps of paper and hearsay, now gives way to the real world of my childhood and its many memories. This world, though some seventy-five years distant, still evokes sensations of touch, hearing, sight and even smell though inevitably the vividness of such memories is tantalisingly elusive; a panorama as it were, shrouded from time to time in swirling mist. Fleeting visions of my mother sweeping the ever invasive dust and grit from our forecourt; my father setting-off for work on his old BSA bicycle, an unbuttoned raincoat flapping in the breeze; the sound of the treadle on my mother's Singer sewing machine as she toiled to keep us all well clothed; the smell and taste of bread pudding and bubble and squeak; all come back to me with amazing clarity. These and many other such memories flood through my senses as I strive to recall those far-off days.

Is it possible to assert that this or that mind picture is one's earliest memory ? If so it may well be that of my mother placing the household steps against the high garden wall which formed the end of our terraced street. From my vantage point at the top of the steps I could just – only just – peer over the wall for a first glimpse of my grandmother, a diminutive figure in black, as she picked her way across the muddy 'Green', a grass covered patch of waste land at the end of the street. Twice a week she made this journey from her near-by lodgings to help my mother for a few hours in the house. The warmth of this memory reflects the intense pleasure which these visits must have given to a child between two and three years of age. My sister Ruth and brother Frank, my seniors by four and five and a half years respectively, were both at school, and these age gaps ensured that until I too went to school at the age of five, I was very much an 'only' child, over-fussed by my mother and much reliant on my own imagination for amusement. Understandably my grandmother's visits must have been high-lights in my week. But before recalling more memories of our family life I will set the scene within the City of Portsmouth where I was born and lived until I was almost seventeen years of age.

Our house, number twenty-eight, was at the far end of Martin Road, Copnor and as such was the last urban housing to the north-east within the confines of Portsea Island. We were fortunate indeed. Not only were we semi-detached with the additional privacy that this bestowed upon us, but we had uninterrupted views over the Corporation Fields to Langstone Harbour a mile or so away in an easterly direction, and over the Green looking north to the high chimneys of the Moneyfields Brickworks. Tucked away behind these were the deep flooded clay pits, forbidden territory when we were young but as we reached the ripe old age of ten plus, armed with penny nets and jam jars, this' Mecca' never failed to yield a catch of sticklebacks, minnows, red throats and the like, to be brought home, fed on ants' eggs and sadly, die an early death. The sense of space and freedom surrounding our house was a boon and a blessing to us children but there was a downside to endure. When the chill Northerlies and Easterlies blew in from Langstone Harbour in the winter rattling our casement windows and freezing the water on the washstands in the bedrooms, our house was probably the coldest in the street but such 'sufferings' were shortlived and stoically endured despite the ever-present chilblains, an all too common complaint in those days of inadequate house heating. Roaming the Corporation Fields and playing in the hay in summer, fishing for green crabs off the sea wall of Langstone Harbour and exploring the reed beds of the old Roman Salterns were more than ample compensation for our winter 'sufferings'.

The road, or more accurately the street in which we lived, consisted of neat though small brick built terraced houses, with tiny bay windows flanked with modest stone pillars, all set behind a slender tiled forecourt separated from the pavement by a low wall and railings. The road itself though edged with kerbstones was surfaced with gravel, to the distress of my mother. Two similar streets, Dover and Folkestone Roads completed our immediate neighboured, nicknamed by us children, Brass Button Alley in recognition of the large number of Petty Officers who emerged from these houses every morning on their way to shore duty in the Portsmouth Naval Base. Due to my parents' Fundamentalist Christian beliefs, the neighbourhood as such played almost no part in our social community life which was centred on the Meeting Room where the Brethren provided the collective support which was needed from time to time. Family life was thus somewhat remote and was lived 'under the eyes of God'. Even now I can see the framed text written in an archaic hand – 'Thou God seest me' – which hung over my bed. As a child I accepted this as a literal truth and many the panics and fears which this induced in me throughout those early and formative years. But what of family life within our neat and tidy home, set apart from our neighbours ?

As I turn to the day-to-day activities of my early childhood these emerge as a series of snapshots rather than a balanced overall picture. Seemingly inconsequential moments – sitting by the fire with my mother, wire toasting fork in hand, toasting a rare treat of buttered crumpets for tea – stand out with clarity, whilst other events of much greater significance are at best half-memories. Whilst our external focal point was the Meeting Room, indoors everything seemed to revolve around the somewhat cramped living room and the large table located in the middle – a table 'for all seasons'. Meals, bible readings, homework, table games, needlework and other occasional activities, all had their 'pecking order' for table space. A book cupboard hung on one wall seemingly by a miracle, but in actuality secured by stout battens affixed tightly above and below. A single casement window looking out over the backyard and garden provided the only natural light, backed-up by the gaslight over the table. How well I remember the ritual of fitting a new mantle. My mother's Singer sewing machine sat under the suspended book cupboard and was in constant use. I can still hear the rhythmical pulsations as the treadle was rocked back and forth by mother as she made a wide variety of clothes – dresses, suits, trousers, blouses and the like – nearly all from cast-off clothing sent by her wealthy sister from London. These were all dubbed as 'Aunty's cut-me-downs' and although not all of them found favour with us children, thanks to my mother's industry we were always warmly and tidily clad. Such diligence took its toll on eyesight and fingers and she suffered much in later life as a consequence.

Two remaining features of the living room need mention and both generate yet further anecdotes. The fireplace, though small in opening, had a wide, tiled hearth and throughout the winter this was stacked with sawn and split ex-railway sleepers, a by-product of my father's job as a railway clerk. These logs were used to supplement the coal supply and were clearly an important contribution to the family budget. Unfortunately they were heavily impregnated with creosote and this gummed-up the chimney which was then swept by my father using an ancient set of rods and a brush. Alas, on one occasion a rod broke leaving the brush securely wedged in the chimney and this necessitated the services of a 'professional' sweep, who arrived grumbling and swearing at the incompetence of 'bloody' dabblers. Such language was not permitted in our house and he was immediately silenced by father, probably accompanied by a relevant quotation from Holy Writ.

The remaining feature was a low door which gave access to the 'cupboard under the stairs'. In addition to holding sundry household items here was housed the all important 'rag bag'. Little went to waste in our family and my mother saved all the cloth scraps from her dressmaking and they were stored in father's old kit-bag, a relic of his days in the Non Combatant Corps in WW1. On rare occasions, as we were finishing our breakfast, my mother would be listening anxiously, as I now realise, for the Rag and Bone man's bell and his raucous call – 'bring out yer rags'. The kit-bag would have already been expertly judged for weight and, either my sister or I would be instructed to run to the ragman and secure 4d for the contents of the kit-bag and then run a further quarter of a mile to the butchers and purchase four pennorth of shin of beef. We would then run the return journey to home wait, no doubt impatiently, whilst my mother checked the meat, and then run to school some half a mile away hopefully arriving in time to avoid the dreaded 'black mark' for lateness. The meat was that day's midday meal. Need I say more ?

The kitchen housed a gas cooker – immaculately clean – a copper, coal fired in my early days, a shallow, yellow sink complete with cold water tap, and a scrubbed white wood table where my mother worked her culinary miracles. Filling and nutritious meals from the most basic and keenly priced ingredients were put on the table day after day and I can never remember any of us going hungry. What a mother – cook, dressmaker, household finance manager, child carer, all rolled into one. A highly intelligent and sensitive person who in today's world would surely have gained a PhD.

But to return to everyday affairs. Hot water was not on tap and was a comparative rarity. Cold water was used for our daily personal toilets save in really cold weather when a big black kettle was filled and heated on the gas stove – to be shared between all three children. Friday bath nights were a strictly controlled ritual based on, to use a modern expression, 'Equality of opportunity'. The 'copper' was filled and lit, a much simplified task when in the fullness of time it was replaced with a galvanised gas-fired version. The bathroom was a tiny room – ex coal store – which led off the kitchen. It housed a white enamel bath – a real luxury in those days – which was equipped with a plug and drainpipe but of taps, either hot or cold, there were none. The hot water from the copper was carried in by bucket, cooled as necessary by cold water similarly carried in and then we children were lined-up ready for 'scrubbing' by one or the other of our parents. Most importantly the sequence was varied week by week on a rota basis so that each in turn had the privilege of the first clean water. The contents of the bath had to serve all three of us though possibly topped-up by additional hot water particularly in winter. The bath also served one other, though rare purpose. In it we children were baptised by one of the 'Elder Brethren'.

Outside was the WC, spacious, clean and whitewashed, with a large gap over and under the door presumably to aid ventilation but also ensuring near freezing conditions in winter. At such times a tiny oil lamp was lit and hung beneath the overhead tank to counter the frost. Toilet rolls were unheard of and in most families cut-up newspaper fulfilled the essential cleaning functions. Not so in our household where once again a spin-off from my father's job as a clerk in the Portsmouth Town Station Goods Office was the supply of books of tissue paper – yet another childhood luxury. These books, some two to three inches thick, were in fact obsolete records – duplicates in copying ink-of invoices and other transactions in respect of goods carried by the Southern Railway in the Portsmouth area. They fascinated me and I would sit reading the contents until my mother, fearful for my health, would enquire in a loud voice as to whether I was still in the land of the living. My standard retort was that 'I was going all the time!'. Small wonder that by the age of four I was adept at mental arithmetic especially in tons, hundredweights, quarters; pounds, shillings and pence, a facility that served me well in later life.

Only one downstairs' room has been omitted from this account – the least used room in the house – the 'front' or 'best' room. In addition to an American organ which dominated the small room, I can only recall an old sofa, re-covered by my mother, some singularly uncomfortable upright chairs and the inevitable aspidistra which occupied a place of honour in the bay window. The American organ or Harmonium as it is sometimes called, came into its own on a Sunday – or in

Brethren's parlance the Lord's Day – after we had all returned from the evening's Gospel preaching. From an early age my sister and I were taught to play this instrument which involved pumping air into the bellows with foot pedals, learning to select the appropriate stops and the ability to read music. It was most suited to the playing of hymn tunes of which my father had a wide knowledge gleaned from many different books.-the Brethren's Tune Book, A and M, Sankey's Songs and Solos and several others. No such musical instrument was permitted within the Meeting Room where all hymns were started 'by ear', hopefully to a convenient pitch for unisonal singing and using the correct metre to fit the words. My father who had a powerful tenor voice also possessed perfect pitch and an awareness of the necessary metre – Long, Short or Special metres etc. – to fit the words. He was generally recognised as our 'ace' hymn starter and from time to time he had the temerity to experiment with 'new' tunes to the bafflement of the congregation! But to return to our home. In addition to the organ I also learnt the violin, with limited success, and as my father was no mean violinist, many a Sunday evening passed with two violins, the organ and my mother singing, all sounding out a wide and variable range of hymns. To this day I find it difficult to walk past an organ be it in Church or Chapel, without testing the lid. Should it perchance be unlocked and a hymn tune book be at hand, I invariably play a tune or two to the surprise and sometimes enjoyment, of any chance visitors. Such was the fabric of our family life, a somewhat broad brush picture which now needs to be filled in with more details of our goings on day by day.

The living room and the all purpose table now return in this narrative, for it was here that virtually all of our family activities were centred. Mealtimes were governed by a set of rules many of which I adhere to this day. Save in exceptional circumstances, the table having been properly and neatly set – table napkins to save soiling our clothes, plates, cutlery and appropriate foodstuffs – we all sat down together, my father at the head of the table. He would then say 'Grace', not the formal and repetitious 'For what we are about to receive...', but an extemporised prayer of thanksgiving for the provision of our daily bread, extended when apposite, for guidance and help throughout the coming day. This was of particular importance at breakfast time – the beginning of a new day – but no meal would be allowed to commence without such a prayer, spoken by my mother in my father's absence until such times as we were old enough to share in this practice. Breakfast consisted of a cereal – frequently Shredded Wheat – followed by brown bread, never white, my father having read an article extolling the virtues of wholemeal flour. Butter could be spread on the first slice but this precluded the addition of marmalade - homemade. Margarine had to be used on all subsequent slices and marmalade or jam could be added in modest amounts in order to cloak the somewhat unpleasant taste of margarine in those far-off days. Cooked breakfasts were unknown – except at Grandma's – although on Sundays and other special occasions boiled eggs were served, much to our enjoyment. This indulgence always preceded one of Dad's walking trips with us children. He asserted that no one could be expected to walk any worthwhile distance without a boiled egg for sustenance. This was linked to another of Dad's principles that at the age of four we children would be expected to walk four miles unaided, at the age of five, five miles, such a progression continuing up to twelve years of age at which point we would all walk twelve miles or even more without difficulty or complaint. Lunch, or dinner as it was known, was taken at midday and was our principal meal. It varied from day to day consisting of stews or braised meat, vegetables in abundance, followed by suet or bread pudding, bread and butter pudding, rice, sago or other equally substantial dishes. Bread pudding is still a favourite of mine whereas pearl barley stew is the only dish which I cannot face! A roast joint was normal fare on Saturdays, consumed cold on Sundays when only a bare minimum of work was countenanced.

Mealtime discipline was strict, good table manners sternly enforced, arguments between us children kept under control and picking at and leaving food not permitted. Indeed any food left on the plate was served-up to the child concerned as a 'prelude' to the next meal and had to be consumed. My mother and father invariably won the 'battle of wills' that often then ensued. It was a harsh but necessary discipline. Waste was unthinkable in circumstances where every penny counted, as the 'rag bag' episode clearly demonstrated.

One meal each day always concluded with a reading of the Bible. Books and Chapters were read in strict rotation and over a long period we would journey from Genesis Chapter 1 to Revelation Chapter 22. Only the inordinately long chapters such as Psalm 119 with its 176 verses, would merit two successive readings and before I left home at almost seventeen years of age this 'journey' had been completed at least twice. From about the age of three we were required to play our part in these readings as they circled the family group, verse by verse. My father would always help us out with difficult words but how he coped with the crippling stutter that I had developed by the age of six, mercifully, I cannot remember. It was a solemn occasion although self-control was tested to the extreme by the erotic language of the Song of Songs, the consequent giggling amongst us children being quickly quelled by Dad. I well remember my brother Frank, five and a half years my senior, having to read 2 Kings, Chapter 18 and Verse 27, the curse of Rab-Shakeh at the time of the Assyrians siege of Jerusalem.

The evident gusto with which he read out the phrase "...and they shall eat their own dung and drink their own piss..." was far from appreciated by my father whose stern glance said it all.

The principle of ' spare the rod and spoil the child ', was generally accepted as a sound principle but applied within our family with restraint. The cane was used but only in cases of persistent or serious disobedience and then never to excess. In fact I have absolutely no memory of childhood resentment on this score. True, we two boys did feel somewhat aggrieved on occasions when my sister, excused corporal punishment, 'got away with' a mere smacking. When my father was not present at a meal, my mother kept the cane by her side as a token of her authority perhaps! A sharp rap from this on the back of my chair sufficed to reduce me to tears though obviously no physical contact was made. The threat was enough.

No mention has been made of teatime, a very simple meal consisting solely of bread and jam followed by homemade cake or sponge. Supper did not exist so tea was the last meal of the day and as soon as it was cleared away and we had helped mother in the kitchen with wiping-up and putting away it was time for homework. Three pieces of yellow oilcloth were unrolled – one for each of us – and with much pencil sucking and head scratching, the various set tasks were completed and submitted to my father for approval. This was by no means automatically forthcoming and many was the time that whole pages had to be re-written as untidiness was not tolerated. In fairness to our fellow-pupils, arithmetical errors were not corrected though we might well be 'invited' to have a second look at some of our answers. Should we need help or advice this was always forthcoming and although my parents were not widely read, they seemed to have a fund of historical and geographical facts at their fingertips.

Newspapers, wireless and the majority of novels did not enter our house though certain fiction – G.A.Henty, R.M.Ballantyne, Fenimore Cooper for example – were approved reading and from an early age I was enthralled by Henty's stories of Mediaeval England and the wars of the British Empire – India and Africa. Strange reading indeed within a pacifist household but they instilled in me a love of history which, to this day, has never left me. My father had a truly prodigious knowledge of the Bible, could give Book, Chapter and usually Verse for any biblical quotation and frequently add the page number from his own personal bible.

Table games were an essential part of our indoor amusements and education. My father would preside over games of skill – draughts at which he was a first-class player – double-nine dominoes played as threes and fives, tiddly – winks and homemade number and word games. Mother would play Ludo and Snakes and Ladders with us, dice games of chance which my father eschewed. No concessions were made by father in order to 'let the children win'. He would teach and help us, but to fiddle any game so that one of us could win was foreign to his nature – and mine. I was adept at dominoes by the age of four but I was at least twelve years of age before I beat Dad at draughts. He was delighted, rushed into the kitchen where my mother was working, to convey the good news. Needless to say I was equally pleased. I knew that I had beaten him 'fair and square'.

Had it not been for my parents unswerving and burning faith, we children could never have enjoyed the fruits of further education, neither would I have acquired the disciplined thinking which characterised my development in adult life. It is only in retrospect that I have become aware of the frugality forced upon them by the necessity to balance the family budget week by week. To get into debt was unthinkable. All bills were paid promptly. Bank accounts and overdrafts formed no part of their lives. When my father was killed in a 'hit and run' raid on Portsmouth on 5th December 1940, my sister, when going through his effects, found the household accounts for the 1920's and 30's. Except on rare occasions there was never more than a 6d balance in hand.

Two of these rare occasions spring to mind and the telling of them will reveal my father as an extremely competent organiser and administrator. The story begins in 1926 at the time of the General Strike. As a Plymouth Brother my father, on grounds of conscience, would in no way feel able to join a Trade Union. Throughout the strike he persisted in attending for work and, with the co-operation of the pickets, used the railway line as a means of access. For some obscure reason which to this very day I do not understand, this action mitigated against his promotion prospects despite his recognised abilities. However, one of the 'top brass' at Waterloo Station, then the HQ of the Southern Railway, was fully aware of his abilities and whenever the Royal Agricultural Show was on the Southern network, he would travel to Portsmouth and personally appoint my father as Agent for the duration of the Show, a period of some two to three weeks spanning the build-up and dispersal periods. The two such occasions concern Swaythling and Windsor Great Park. At that time the principal means of transport in and out of such Shows was of course the railway, and for the enormously prestigious Royal Shows the volume of such traffic necessitated the building of extensive temporary sidings. This entailed not only a feat of engineering, but also the need to establish a comprehensive and adequate administrative organisation. This my father achieved. Except for Sundays he would be away from home for the two or three week period of the Show and his accommodation was always George Vth's ex Royal Coach – luxury indeed. The extra pay attached to this appointment must have been a God-send in a very real sense, not to be frittered away on current expenditure in the way of luxuries and treats, but used, I have no doubt, for home improvements and probably to pay off a small house purchase loan from a well-to-do Plymouth Brother in our local Meeting.

Sickness and illness of varying severity occurred from time to time. 'Run of the mill' problems – colds, tummy upsets and the like – were treated by my father who, in common with a majority of Brethren, believed in and practised Homeopathy. Homeopathic medicines were available from one of the Brethren who I presume was a qualified Dispenser. Aconite, Belladonna, Arnica, Bryonia, Nux Vomica and many more such medicines, were all regular inhabitants of our family medicine chest and at the first signs of a cold a few drops of this or that – frequently bitter to the taste – were added to a tumbler of water and taken in accordance with my father's instructions. Did they cure ? I know not. Dad was aided in his diagnosis and prescribing by reference to two books – Dr Durden's and Clarke's Prescriber. More serious problems meant calling in

Dr Hall, a homeopathic doctor from Southsea. He would cycle all the way from Southsea to Copnor whatever the weather, treat the patient and then charge only what he thought my parents could afford. This was his normal approach to everyone of his patients, sometimes visiting and prescribing 'for free'. A real saint. My mother's health always seemed precarious, sometimes necessitating a prolonged stay in hospital. On such occasions we children were cared for within the tight-knit Brethren community who were always fully supportive of those in trouble, distress or need. This was a positive aspect of their 'Exclusivity'.

The picture I have endeavoured to paint of our simple and at times restrictive family life, does it less than justice. Despite the restraints, many of which only became apparent as we reached our teens, there was an enormous wealth of love and affection which surrounded and supported us at all times. Sadly, it is only in retrospect that one becomes aware of the parental self-denial and dedication which underpinned this childhood of mine.

So enough of the family life within our trim and tidy house. As a matter of course much of our time was spent out-of-doors, especially in the summer months when the near-by open spaces beckoned. In addition to making hay in the Corporation Fields and fishing in the flooded clay-pits, already referred to, we rambled as far as Langstone Harbour, perching on the sea wall fishing for crabs with bent pins and bread and more daringly exploring the reed beds in the old Roman Salterns. None of this rated as walking in the countryside, a treat that had to await one of my father's rural expeditions. Sometimes a whole day would be devoted to our local wanderings and then a bag of homemade buns and a bottle of lemonade would provide the excitement of a picnic. Closer to home the streets were the natural playground for the children of the neighbourhood, perfectly safe, as private motor cars were virtually non-existent. Almost without exception, tradesmen carried out their deliveries by horse-drawn vans and carts.

Fortunately, at this time my parents did not enforce too strict a control on our friendships outside the Brethren, providing that such friends passed muster as respectable and well-behaved. I joined such friends in the games of those times – hopscotch, fives, whipping tops – turnips, carrots, windowbreakers and such like – but team games I avoided as much as possible. I was during those early years small and far from robust. An iron hoop complete with a hook, and a 'wheel' were my favourite toys. The latter, probably an old pram wheel retrieved from a nearby tip by father, had a block of wood driven into the hub and then a large nail hammered into the wood thus providing as it were, a short axle. With the aid of a suitable stick I would 'run the wheel' for miles round the streets, up and down kerbs, round lamp posts and many another trick as I became more and more expert. Such a toy cost nothing and had the great advantage as far as I was concerned, that it could be enjoyed in total isolation.

Mention has been made of our gravelled roads, which in dry weather were a nightmare for the industrious housewife of whom my mother was a leading example. Dust and grit blew hither and thither in the wind getting in eyes and noses and penetrating ill-fitting casement windows and doors. At such times the Corporation Water Cart was awaited with barely disguised impatience and a collective sigh of relief was heard when it at last arrived. Was it a Portsmouth speciality ? At the end of our street was an odd looking hydrant, a tall vertical pipe bent over at right-angles at the top. The Water Cart would be placed under the protruding pipe, somewhere a valve would be opened and lo and behold sea water or brine, would gush forth and fill the tank of the cart. To the rear of the cart was a 'rose' some six feet wide and as the old horse plodded along the road salt water soaked the gravel and laid the offending dust. Apparently this sea water main extended round much of Portsmouth as it then was. Even more unusual was the Water Tram which was housed in the North End Depot. At times of extended dry weather and heat, the wooden blocks in which the tram lines were laid, would shrink with somewhat obvious consequences for the safety of the trams. Before this could happen, the Water Tram would be filled-up, I presume with brine, and it would trundle forth over the tram network, spraying as it went, soaking and swelling the wooden blocks. This was such a rare sight that when warned by the 'jungle telegraph' that the monster was out and about, we schoolboys would take up strategic positions in the hope of seeing it pass by.

The tram and the bicycle were, at this time, the principal means of transport for the 'working' population amongst whom car ownership was but a pipe-dream. As a consequence, each working day, thousands of 'dockies' cycled to and from the Naval Dockyard, the main local employer up to and throughout World War 2. This created the famous 'Dockyard Rush' a sight which would have provided a tourist attraction in this day and age. As the Dockyard hooter blew for the midday one hour break the sound was heard all over Portsmouth. A solid phalanx of hundreds upon hundreds of cyclists poured out of the Unicorn Gate and headed for Commercial Road. Extra police were on duty and oncoming traffic was brought to a halt as this tidal flow of 'wheeled humanity', all desperate to get home for lunch, swept all before it until it gradually thinned out, left and right to their several destinations. What a sight, which was repeated in the evening and on every working day of the week. What if they had all been in cars? Doesn't bear thinking about! All this excitement passed us by and my father made certain that his lunch times did not coincide with the dreaded 'Rush'.

Portsmouth as such had little appeal to us as a family apart from essential shopping trips, though I remember going with my mother to McIlroys, a drapery and clothing store, and watching with fascination as the shop assistant, having listed our purchases on a slip of paper, placed it in one half of a wooden sphere, screwed the other half on and then elevated it onto an overhead track. From there it rolled down a decline to the central cashier's desk, was costed and re-elevated, returned to the counter assistant who took the money and again returned it to the cashier. The shop was criss-crossed with these elevated tracks loaded with wooden spheres rolling back and forth to the central desk – much more interesting than an ordinary till. Only two other local family trips are worthy of note. Sometimes my mother who hated the beach and suffered if sunburnt, would pack a quaint collapsible wooden pushchair with a picnic and we would set-off for Fort Cumberland beach, the only sandy beach in the vicinity. Its disadvantage was the proximity of massive naval gun turrets on the land just behind the beach and on practice days these would be firing on targets far out to sea, the shells 'whooshing' over our heads as we sat on the sands. This was definitely not to my mother's liking. Of much greater appeal to her were the newly created illuminated Southsea Rock Gardens which before the War where both innovative and a great crowd-puller. A short distance away was Clarence Pier with its fun fair and here, to my mother's almost girlish delight, we saw our first Bumper Cars – a sight that never failed to amuse her.

Yet another by-product of my father's job was an annual allocation of free railway passes for the family and an unlimited number of privilege tickets at one quarter or one eighth full fare. Although holidays away were not possible, day trips using such tickets were a regular part of our lives, particular in summer. My sister and I would go over to the Isle of Wight by paddle steamer and then travel around by train to whichever bay took our fancy. Trips with Dad took us as far afield as London, Oxford, Exeter – by the Atlantic Coast Express – and shorter trips to Midhurst, Cocking and other little stations in Hampshire and Sussex whence we would set-out on one of Dad's favourite walks. The Meon Valley line was used to visit my Grandparents when they were living in West Meon. To be left there for a few days stay was heaven indeed. My Grandfather – ex-Colour Sergeant – would have us on CO's Inspection every morning. Boots, with their laces removed, had to be polished even in the in-step, finger nails clean, hair brushed and parted and generally our turnout had to be spick and span. He was superficially a fierce old man with a waxed moustache but utterly lovable and adored by us children. But above all Dad's favourite haunt was the New Forest and whenever possible we caught a train to Lyndhurst Road Station – now re-named Ashurst – and with picnic in pushchair walked to Busketts Lawn where we established 'camp' before wandering through the deserted Forest. One day who should we meet but the 'snake man', Brusher Mills carrying a canvas bag. He showed us all the contents, several adders which he tipped out onto the grass and then collected with something akin to a pair of crucible tongs.

Having done this he then lectured my father on the dangers of allowing his children to wander through the bracken in sandals and short socks. What a privilege these railway trips were for us children. It not only broadened our horizons, but by the age of fourteen I was a hardened rail enthusiast, travelling hundreds of miles by myself and already expert at deciphering the mysteries of railway timetables.

Come the early thirties our local environment underwent a considerable change. The 'Green' at the end of our road, together with adjoining fields and allotments, was developed as a housing estate by a Portsmouth builder – Councillor Winnicott- under a scheme known as Subsidy Housing, the financial details of which I have no knowledge. Semi-detached houses were built to a standard specification – three up and three down – and the builder was able to extract sufficient clay from the site with which to build these houses. We were thus able to view the hand-making of bricks at close quarters from puddling to kilning and occasionally were allowed to 'help'. Before 'pegging-out' for the foundations got underway, my father had the foresight to approach the Builder for a strip of land alongside our end-of-the terrace house and a six foot width was purchased for £6, A boarded fence was erected, the existing garden wall demolished by father and the bricks, cleaned of mortar by my sister, used to build a workshop and coal store at the end of our garden. The side entrance thus created was a great improvement and certainly added to the value of the house.

Within the next two years my Grandfather died and as my Grandmother's wealthy daughter and well-to-do son declined to help, my father and mother accepted that she would have to come and live with us. This necessitated a move to a larger house and we duly left Portsea Island and moved to mainland

Cosham to a new end-of-terrace house in a cul-de-sac, Park Grove. The garden overlooked a Church playing field with tennis courts and a croquet lawn, both of which gave us endless and friendly entertainment. I have often wondered how my father funded this move, but his recent promotion to Goods' Agent at Fratton Good's Yard must have helped considerably, plus the fact that my elder brother had by this time left home and was self-supporting. Portsdown Hill was but a 'stone's throw away', on its far side glorious countryside, farms for fresh eggs, copses for bird-nesting and quiet lanes for walking – joy indeed. School was now some two miles distant and nominally a trolley-bus ride away but by running this distance each way I could enhance my pocket money!

The house, compared with number twenty-eight Martin Road, was modern and spacious and our immediate environment greatly improved, but in other matters little had changed until my sister, a constant companion of mine, left for Reading University a year after our move. Our earlier financial hardships had eased considerably and when a year later my Grandmother died of a stroke and with only me at home, frugality as we had known it was a thing of the past and my life was far less governed by pinch-penny disciplines than heretofore.

Despite this improvement, when my sister as a result of developing an appendicitis lost her State Bursary at Reading University, my father could not bridge the financial gap thus created and, tragically, she was forced to terminate her Degree Course and return home to train as an Elementary School teacher. She was never able to accept this blatant injustice which frustrated her legitimate and achievable ambitions. My own educational achievements which can but be described as 'middling', suffered no such trauma, but such as they were follow on in a chapter of the utmost brevity. Meanwhile I draw the line under this fragmented and incomplete chronicle of my childhood at best described as 'snapshots through a swirling mist'.

CHAPTER 4

In the Shadows

Shortly before my fifth birthday, my mother took me to George Street Elementary School, a drab building, overlooking a vast Municipal Cemetery on one side and a cluster of mean streets on the other. The trams rattled by on their way to the centre of Portsmouth. I was taken in to meet Miss Tranter the mistress of the 'Baby Class' who was told by my mother that on no account was I to play with plasticene – nasty, messy stuff which could soil my clothes! Thereafter, for the duration of my time in the Infants' School, I joined the girls for knitting! To this very day I still resent this humiliating start to my school life. The basic 3 R's had been instilled in me at home and when Miss Tranter went through the ritual of testing and assessing the 'newcomer' she was undoubtedly surprised when I rattled-off my twice to twelve-times tables, read the 'baby' books with no difficulty and copied various words in readable print. I was immediately elevated to Class 2 and to my regret missed the joys of playing on the sand-table, building with coloured wooden blocks and other equally enjoyable and non-demanding games.

There is all too little to recall of the schooling of those days. Classrooms were strictly functional, ill-heated in winter, learning was by rote and discipline, based on corporal punishment, varied from teacher to teacher, many of whom were over-lavish in the use of the cane. Whilst still in the Infants' School I was selected to stand-up in front of the whole school to recite a poem. Alas, 'pride cometh before a fall', and by the time I was six or maybe, seven, I was afflicted by a crippling stammer, not so acute when at home but sufficient to completely silence me elsewhere. Even by the age of seventeen I was unable to ask for a destination on tram or bus and could but tender the exact fare for the required trip. The origins of this stammer are somewhat obscure though I grew up with the idea that it was a direct consequence of being pinned against the wall at the top of Copnor Bridge by an errant motorbike and sidecar which failed to negotiated the 'S' bend and mounted the pavement. Looking back I am of the opinion that the incident was only part of the story – a trigger perhaps. 'Over-mothering' undoubtedly contributed to this and other childhood problems, many of which were only resolved when I left home in 1937.

The mention of Copnor Bridge brings back memories of minor adventures on the half-mile walk to and fro to school every day. The bridge crosses the main Portsmouth to Waterloo Railway Line and is quite steep and winding. Horses and carts and Foden Steam lorries wound their ways up and over the bridge, the former sometimes having to use the services of a 'lead' horse, stabled nearby, in order to get a heavy load over the bridge. We kids would sometimes dash under the cart and sit on the rear axles for a free ride. At this point, children on the pavement would shout out in chorus 'Look be'ind governor ',the carter would with great expertise flip his long whip back over his shoulder, causing the 'free-riders' to dismount instantly. The tram lines terminated at the foot of the bridge from where a frequent service ran into the city. Half-pennies placed on the line just before a tram passed emerged flattened and enlarged to be used in due course in certain slot machines, though rarely with any positive results. Such harmless fun enlivened

The author.

otherwise dull days at school where my transition from Infants to Junior at least freed me from the ignominy of the dreaded knitting.

I progressed satisfactorily through Junior School though with no great distinction. My early promise seemed to have faded. At one stage I sat alongside a cousin whose father – my own father's elder brother – was our teacher. At the commencement of the school year we were introduced to the rest of the class and a wholly unjust punishment 'tariff' announced. When two strokes of the cane were deemed to be deserved by other pupils, four strokes would be my quota and six for my cousin – just to show that there was to be no favouritism! Only two other fragments of memory come back to me from this period of my education. My uncle striding up and down in front of the class intoning Henry Newbolt's, 'Vitai Lampada' – There's a breathless hush in the Close tonight...' and the school foregathering for Empire Day when a very large map would be unrolled showing in red all the possessions within the Empire 'on which the sun never sets'.

In due course I had to sit the highly competitive entrance examination for the Portsmouth Northern Secondary School. Some 1200 or so entrants competed for 240 places, but in order to secure a scholarship, an essential requirement for reasons of domestic financial stringency, the higher the placing in the results the better, and certainly a position in the top sixty was vital as this guaranteed entry to the 'A' Form. Both my brother – brilliant – and my sister – dedicated – had, in their day finished in the top thirty. My 68th position found little favour with my father but also placed a scholarship in jeopardy. Luckily for me due to the withdrawal of other applicants, I just scraped into 60th place and the desired 'A' Form. My cousin and fellow-pupil failed, much to my disappointment, as he then went to the fee paying Grammar school and our close association was interrupted..

My new school – Portsmouth Northern Secondary – was indeed 'new' as it was opened in 1932, the year I joined. Even by today's standards it is still an imposing building. In 1932 it was rightly judged 'magnificent'. The central hall, capable of holding 1,000 people, was complete with a stage which would have done justice to many a Provincial Theatre, with orchestra pit, dressing rooms, spacious balcony and exciting decor. On each side of this central hall, built around a quadrangle, was the boys' school to the East and the girls' to the West (never the twain shall meet!). The form rooms, centrally heated, were simple but well equipped as were the laboratories and gymnasium. Metal and wood workshops provided for craft training. In those days it was an educational establishment to be envied and I was indeed fortunate to attend there for my Higher Education. Despite these excellent facilities my school career was singularly undistinguished as I alternated between 'A' and 'B' Forms, to the despair of my father. My French master, under whom I had the misfortune to study for four years, all but destroyed my self-confidence by exposing my speech impediment to ridicule in front of the class. Small wonder I only scraped a Pass in French in my Oxford School Certificate and then only thanks to my sister's encouragement and coaching at home. I excelled in History despite my refusal to learn scores of 'dates' and secured a special recommendation in my School Certificate. Of all my masters the Biology teacher, Mr Walker, undoubtedly contributed most to my subsequent development. In marked contrast to my French master he empathized with my speech problem and in addition to appointing me as custodian of the school's aquaria, he encouraged and eventually persuaded me to give a lecture on this subject to the Biology Society. He was convinced that if I gave an epidiascope talk in the dark I would be more relaxed and freed of my stammer. Bless him, he was right and from then on I realised it was curable, though a further two to three years were to elapse before I all but completely laid this particular ghost.

In addition to my mediocrity in the academic field, I fared no better where sport was concerned. I was puny – less than five feet tall at the age of sixteen – and timorous. Rugby and cricket scared the wits out of me and I avoided them whenever possible, but one accomplishment rescued me from total anonymity. I could run considerable distances, up to ten miles, at a 'competitive' speed! In part this arose from my habit of running two miles each way to school every day, subject to weather, saving the fare and, with my father's agreement, adding it to my pocket money. This must have built up my stamina, with the result that I excelled at cross-country running and was selected to represent the school in the South of England Championship. I performed creditably, the high-point of an otherwise lack-lustre athletics record.

Meanwhile my brother and sister had entered University – the former, who was brilliant, failed owing to indolence and the latter who was diligent was terminated due to ill-health. Talented cousins overshadowed my lack of achievement. One of them, my junior by some two or three years, was a true genius, a brilliant mathematician and physicist, a PhD at a very early age and likewise one of the youngest ever Fellows of the Royal Society. He was one of the leading nuclear physicists of his generation. Such family talents did little for my self-confidence. It was some forty years later that the sole survivor of my father's generation told my wife and me that I had always been considered the 'bright one' of my family by her late husband, my Uncle Ernest, himself a highly talented and much 'decorated' Admiralty expert on torpedoes, mines and rocketry. Pity I wasn't aware of this whilst I was at school, or possibly I was a super-late developer!

Despite securing seven Credits in my Oxford School Certificate – but only a Pass in French – a combination of circumstances terminated my formal education. By our family's standards I was a mediocre academic; my brother's and sister's University problems had drained my father's meagre financial resources; my mother's precarious health was causing fresh anxieties, though in the event she lived for another thirty-two years. In 1937 at the age of sixteen I sat the Civil Service Clerical Entrance Examination – passed – though not with flying colours, and was in due course appointed as a Treasury Grade Clerk in the Directorate of Naval Accounts, Branch 1, (DNA 1 for short) in Whitehall, London. Whilst awaiting the results, my father also entered me for the London County Council Clerkships Examination and together with my cousin of Elementary School days, we journeyed to London to 'sit' the exam. It lasted for two days and during that time by devious means – unknown to our parents – we spent an afternoon at the London Palladium seeing and enjoying the Crazy Gang Show, my first introduction to the world of entertainment. It showed me a world beyond my wildest dreams and was doubtless a highly 'educational' experience, due almost entirely to my cousin's foresight and initiative. That I passed the exam is irrelevant as the Civil Service result came through in the meantime and was clearly the preferred choice.

My mother was anxious to keep me at home and before the Civil Service appointment was formally accepted, my father, through the good offices of a friend, a Plymouth Brother of some standing, arranged for me to be interviewed with a view to joining the National Provincial Bank in Portsmouth. I passed the interview but was rejected as I was only 5 feet tall and a minimum height of 5 feet and 2 inches was required! Maybe I couldn't have seen over the counter! This bizarre rule was to have profound and beneficial effects on my life. Is there in fact a 'divinity that shapes our ends rough-hew them how we will'? On 19th July 1937 some three months before my 17th birthday I left home for London, never to return save for the occasional visit.

CHAPTER 5

God or Mammon

I could not have known that the non-return railway ticket from Portsmouth to Waterloo Station was to be the beginning of the most decisive period of my life, stretching from the 19th July 1937, when I left my home, to the 25th November 1940, when I joined the Army in World War Two.

On that July morning my father saw me off from Portsmouth Town Railway Station on a trip that I had done many times in his company. The recently electrified line provided an hourly express service to London in one and a half hours and an early train ensured that I would be at Waterloo Station in good time to be met by my Uncle Ernest at 9.30 or thereabouts. As far as I can recall I had but one large suitcase containing the whole of my wardrobe, new clothes which my parents had purchased at the Co-op Clothing Store in readiness for my living away from home. The grey and blue warm dressing-gown, bearing the Co-op label – 'Labour and Wait' – is still in use. I cannot pretend that this first day is etched into my memory but one or two episodes come to mind. Uncle Ernest has already been referred to and he it was who worked in the Department of Torpedoes and Mines in the Admiralty in a senior technical-design role. He was, as was the rest of my father's family, a Plymouth Brother, and had offered not only to introduce me to my Departmental Head but also to arrange suitable lodgings for me In London within a Brethren family.

Waterloo was undoubtedly reached 'on time'. I probably staggered down the platform with my heavy case – all five foot and half-an-inch of me complete with grey trilby hat – to find Uncle awaiting my arrival. Incidentally headwear for men was an essential part of the Brethren's uniform', on the principle that when entering a Meeting Room, the removal of the hat was an indication that you were 'praising God'. I will also add at this juncture that when in due course I could afford it, I changed to the in-fashion 'pork-pie'. But back to more important matters. We crossed London, walked up Whitehall, passed the Horseguards on sentry duty and entered the Admiralty where two doormen, each bedecked in an ancient uniform, sat in large canopied seats. Up we went to my Uncle's office in 'Old Bank Building' where my case was deposited and then wound our way down many a staircase into the maze of underground passages which ran beneath the Admiralty buildings. In WW2 these were converted into Operation Rooms, Air Raid shelters for the Government and accommodation for Winston Churchill. We then climbed yet more stairs and arrived at the Directorate of Navy Accounts, Branch 1 (DNA 1) where I met both the Departmental and Branch Heads, a somewhat nerve-racking experience for a small sixteen year old boy – for a boy I was still. Never again did I darken the 'holy of holies' to talk to Max Robinson the head of all the DNA's.

The Head of Branch 1, Mr Skinner, was a very serious but fair-minded man who in due course had to cope with the arrival of three young teen-age clerks, the first such intake into the Department since 1921. That we all survived the experience is a tribute to his tolerance. I have however moved several weeks ahead in anticipating the arrival of Cyril Pickering from North

London and Jack Jonas from Rattery in Devon, who were to become close friends of mine. To return to my first morning. Uncle departed leaving me with Mr Skinner who introduced me to the Section where I was to work – the Naval Pensions Records – staffed by two elderly ex-WW1 veterans and a somewhat younger- probably forty plus years of age – senior clerk in charge. Arrangements had been made for Uncle to come over to meet me for lunch and this resulted in the same underground trek, to and fro, which would be repeated at the end of the working day. I was taken for lunch into the Dining Room, used mainly by senior Civil Servants where a 3 course lunch cost in the region of 1/6d. Next door was a cafeteria for the more lowly paid where an adequate meal averaged 8d or so. Access to these facilities was secured by a Luncheon Club Card, issued free and supposedly non-transferable! Until our particular office moved out of Whitehall, Uncle treated me once a week to a posh meal in the Dining Room and in so doing not only saved me precious pennies but also introduced me to the intricacies of dealing with more 'exotic' foods than had previously ever come my way. He did much to shepherd me through my early days away from home. This memory lingers still. At the end of the working day one of the clerks took me, once again underground, to meet Uncle, my case was collected and a Number 10 bus was boarded for Peckham Rye, SE15, where in nearby Evelina Road I was to lodge with a Plymouth Brethren household, behind and over the small Gentlemen's Outfitters which they owned. The charge was 21/- per week which covered half-board Mondays to Fridays, full-board at week-ends and my personal laundry. Nothing was said as to my obligation to work in the shop from time to time during evenings and Saturdays! I was determined to be self-supporting and to the best of my knowledge I never at any time wrote home for money. My salary was £75 per year – 28 shillings and 8 pence per week so after paying for my 'digs', I was left with about 7 shillings for all other purposes. Bus fares and lunches accounted for between five and six shillings a week so little was left for contingencies. Fortunately I had only to 'survive' until my seventeenth birthday on the 6th October when my salary rose to £85 per annum, an increase of 4/- per week, and I even began saving money in the office Loan Club, an excellent Institution run expertly by one Freddy Matthews. Before returning to life at the 'Office', my experiences in my new 'home' need to be chronicled.

I was still under the disciplines of the Brethren and found myself coerced into attending a full quota of weekly meetings – three times on a Sunday plus Monday, Tuesday and Thursday evening. As I was not as yet 'Received into Fellowship for the Breaking of Bread', I was not permitted to attend the Friday evening meetings which dealt with business and doctrinal matters. What a relief! On Sunday mornings now that I was detached from my family I was required to 'Sit behind the Board' thus ensuring that I could not participate 'illegally' in the Breaking of Bread'. Mercifully the Board was large enough to conceal a book so there I sat reading some Novel that I had managed to borrow and smuggle in. Thus began an ever-increasing series of deceptions and, eventually and regrettably, hypocrisies which characterised the dichotomy in my life throughout the succeeding three years. Added to these problems which oppressed me, I undoubtedly suffered severely with 'home-sickness', exacerbated by the totally urban environment of SE15 and the sub-standard cooking of my pinch-penny landlady. The final straw was a dish of pig's trotter with the muck of the stye still present in the folds of the skin! My refusal to eat this dish and my more than outspoken criticism of her cooking was the end of a less than happy relationship. But I have moved far ahead of my first day in London and my subsequent 'Office ' experiences so must retrace my steps to my second day, unaccompanied by my Uncle who lived far-away in Croydon.

Office hours were flexible – 9 to 4; 9.30 to 4.30 or 10 to 5, the latter being the preferred option for most of my fellow Civil Servants and the one I adopted until I left SE15. I boarded the No 10 bus in good time, tendered my sixpence (I still suffered with a crippling stammer) and set-off for Whitehall where I dismounted. But where did I work ? I had only approached the Accounts' Department by underground passages and had no idea of its location. Whitehall is a large and imposing array of offices and my wanderings on that increasingly fraught morning took me to the Foreign Office, Board of Trade and other entrances without success, until some kindly messenger directed me to a small street to the North of the Mall, Spring Gardens off Cockspur Street, where I finally arrived late and agitated for my first full day's work. The story told, I was excused amidst a degree of amusement from my fellow clerks who, there and then, decided I needed to be inducted into the mysteries and sights of Whitehall and its immediate environs. This was a process that took place from time to time over the coming weeks but only two of these experiences stick in my mind. The first was to walk down Whitehall and report back as to the number, location and names of the pubs. In those days no less than four pubs existed cheek by jowl with one another, one of which was called 'The Shades'. I wonder if it is still there? The other test of my powers of observation for which pouring rain was a prerequisite, required me to go and stand on the steps of St Martin's-in-the-Fields and look upwards at Lord Nelson on the top of his column. Due to the demands of suitable weather this led to me leaving the office during working hours! What a sight! The pouring rain cascading down Nelson's statue, ran into the hilt of his sword and from a precise angle it looked all the world as if he had unbuttoned his flies and was indulging in a 'pee'. I am informed that the Ministry of Works has long-since blocked the groove in the hilt of the sword and thus denied Londoners and visitors this somewhat unusual attraction. However, office life was by no means all amusement; indeed the card index records I was assigned to were intensely boring, futile and certainly did not accord with my anticipations. After a few weeks I plucked-up courage and asked for an interview with Mr Skinner the Head of Branch. I had the temerity to tell him that the work I was doing was not to Treasury Grade Clerk standards and could I have a transfer ? 'Request granted' took me by surprise and I was transferred to the Section where the final accounts books were compiled for all Naval expenditure, there being a separate ledger for Receipts and Expenditure, posted on massive lecterns in immaculate copper-plate by two very experienced clerks. The compilation of these final accounts was by way of many subsidiary ledgers, each apportioned to a specific class of expenditure – ships, hospitals, buildings, administration etc – classified as Votes which in turn reflected Parliamentary approval for the Naval Estimates as laid before them. In short, Parliament 'Voted' or in other words, approved Naval Expenditure on an Annual basis and this was enshrined in the Annual Book of Estimates – the DNA 1's Bible. Such was the detail that there was even a Vote sub-heading for spending on 'Rubber bowls – Officers Insane for the use of'.

My own limited responsibilities in my new job was to keep a subsidiary ledger entitled Vote 10, New Works, this being a collation of all new buildings, extensions and repairs for the Navy, worldwide. Not as exciting as it sounds but a great improvement on Pension Card Index Records.

It was at this time that Cyril and Jack arrived in DNA to complete the 1937 intake, the first since 1921. It must have been something of a culture shock for the staid Civil Servants who had been left undisturbed for sixteen years, but not only were we tolerated but also humoured as 'The Three Young Musketeers'. Increasingly we became the closest of friends and during lunch-times and after work I spent more and more time in their company. We would deny ourselves cafeteria lunches and make-do with 2d ham rolls. The money thus saved was spent in a variety of ways. On rare occasions we would have lunch in the Strand Lyon's Corner House – elegance indeed – for 1/6d ; more daringly an equally occasional visit to The Windmill, then the most daring 'Show in Town'. I dread to think the repercussions that would have ensued had the Brethren had an inkling of these secret and depraved activities. Lunch-time visits to Soho were a never-failing source of amusement, shops selling exotic foodstuffs, street markets and always the prostitute on the street corner who exchanged pleasantries with the young men who were clearly not potential clients. At least once a week a few minutes looking-in at the Irving Statue behind the National Portrait Gallery, was bound to be rewarded with typical street entertainers. Flame eaters, sword swallowers, card sharpers, the ex-Inspector of Police who laid claim to the Throne of England and many more such characters took their stand there day after day relying on the pennies thrown 'into the ring' for much of their income. Sadly all of these lunchtime forays came to an abrupt end when we were transferred, lock, stock and barrel to Great Westminster House, Horseferry Road, Westminster, a modern office block which at least had the merit of releasing us from dreadfully cramped conditions. I suspect it also had more than a little to do with the ever-increasing preparation for war.

The Munich Pact of September 1938 brought changes to our personal lives. Cyril Pickering and Jack Jonas both enlisted in the Territorial Army, the former in the Queen Victoria's Rifles and the latter in an Anti-Tank Regiment. Still entrapped within the Brethren, this option was denied to me as an obligatory pacifist. However, I trained as an Air Raid Warden and at the time of the Munich crisis spent time in Lambeth fitting civilians with gas masks and explaining their use in the event of gas warfare.

At about the same time as we moved offices I moved to new lodgings in Goodmayes, Essex, a far more congenial environment with open countryside near at hand. My new landlady, a widowed Plymouth Sister with her two spinster daughters, was a motherly soul and an excellent cook. A great improvement on SE 15. I was now 13 miles from work so with the aid of the office Loan Club I purchased a James 3-speed bicycle, cycled to work and repaid the loan from the money saved on fares. I now decided to make a determined effort to rid myself of my stammer which was still a hindrance and embarrassment. Working on the principle that nobody stutters when they sing, I would go for long walks on dark winter evenings, singing all the phrases which had troubled me in the course of the day, then repeating them again and again in normal speech until such time as I felt confident enough to use them in everyday conversation. I walked many miles for hours on end, singing and speaking to myself – thankfully nobody listening – and finally all but conquered my impediment, though on rare occasions certain words still 'catch me out'.

Despite my move to Goodmayes, nothing changed in my world of Brethrenism except that I now attended the local Meeting Room where, shortly after my arrival, a petite and vivacious seventeen year old girl appeared as if from nowhere. Although not Welsh by birth, she had come from Pontypridd to live with an uncle and aunt on the initiative of her parents, who sought to distance her from a boy friend whose only 'crime' was non-membership of the Plymouth Brethren! Thus I met my wife-to-be, the most momentous and fulfilling happening of my life. Her uncle, Frank Seaman, was the most dogmatic, narrow-minded and dictatorial Plymouth Brother that it was ever my misfortune to meet and his attitude to our 'courtship' was a major contributory factor to an act of hypocrisy which had 'inevitability' written all over it. As neither Hilda Campbell nor myself had been received into Fellowship, every conceivable opportunity was taken to destroy our growing friendship, though fortunately as we both worked in London, lunch time meetings were unaffected. I had acquired a tandem but even sharing this with Hilda was banned. Finally in desperation we decided that the only way out of the impasse was for both of us to ask to be 'received into Fellowship', thus sanctifying our courtship! The relevant rituals were put in train and we were both individually examined by 'Elder Brothers'. Hilda was accepted 'as one weak in the Faith' whilst after four hours of rigorous examination it was decided that I would have to 'wait on the Lord' and be re-examined. Clearly these Elders were not fooled by my well-prepared answers and realised that my heart was not where my head was. Nevertheless, some four weeks later I was accepted into Fellowship and Hilda and I could openly enjoy the pleasures of riding a tandem in the Essex countryside!!

I had by now reached eighteen years of age, was earning the princely sum of £104 a year and had been promoted to train as an Auditor of ship and dockyard accounts which had necessitated a move from DNA 1 to DNA 2. The work was more demanding but of far greater interest, particularly as I was handling overseas accounts which were shipped from the outposts of the British Empire in well constructed packing cases bulging with documents of all shapes and sizes and in a multiplicity of foreign currencies. Sorting out this paperwork into organised accounts was far removed from the routines of DNA 1 but mentally more satisfying.

As 1938 gave way to 1939, talk within the Admiralty was more and more concerned with the very real prospects of war with Germany and our internal ARP – Air Raid Precautions-organisation was up-dated and re-organised. I was still committed to Warden's duties in the civilian community and purchased a 125cc Excelsior 2-stroke motorbike. I was granted a special driving licence 'for maintaining services essential to the life of the community' but as September 1939 was reached events overtook us all, completely changing the pattern of our well ordered lives. Cyril and Jack were ordered to join their Territorial Units which by now were being placed on a war footing; Hilda returned to her family in South Wales; the Directorate of Navy Accounts was preparing for evacuation to Bath and I, together with a more senior clerk, was instructed to form the 'rear party' to supervise the loading of all our records onto a fleet of pantechnicons and ensure that nothing was left behind. This duty completed we both motorcycled off to Bath where I arrived 'on tow' having run out of fuel some miles short of our destination.

The move to Bath and the associated direction into Government organised lodgings, was yet another defining moment in my individual development. For the first time in my life I lived with non-Plymouth Brethren as six of us, all Admiralty Civil Servants, shared the same lodgings. Bath, the city of Regency terraces and magnificent streets, the Abbey, the Roman Baths and Assembly Rooms, was home to countless ex 'Colonials' and retired senior officers from all the Services, and the arrival of the Admiralty with its attendant 'horde' of Civil Servants was far from welcomed. The local Press made this abundantly clear.

Our own requisitioned lodgings were ruled by a hostile landlady who had known better times and ere long four of us found alternative accommodation on the outskirts of Bath with a country landlady whose aim in life was to serve us up enormous meals, well cooked, till at last we failed to clear our plates. A delighted Mrs Oborne shouted out to her husband 'I've beaten 'em at last Sid!' A lovely buxom, friendly person who treated us all as members of her family. She wept on my shoulder when eventually I left to join the Army.'

Sharing not only office life but also home-life with these new friends opened up a completely new world for me. One of them, a dedicated Christian, persuaded me to join the Civil Service Christian Union (CSCU) although such an association was contrary to the basic tenets of the 'Exclusives'. Meeting with this essentially ecumenical group taught me that the Exclusives were by no means the sole recipients of the Truth, in itself is an elusive concept which comes in a variety of 'packages', all valid in the eyes of other Christian bodies. On a less elevated plane, I came into contact with Playing Cards – totally banned in my home – and was taught the rudiments of Contract Bridge which led to an extraordinary evening out at the prestigious Assembly Rooms. I needed no persuasion to go with two of our number to a Bridge Drive, for the first and the last time in my life. I knew absolutely nothing of the complexities of the game and 'conventions' was a foreign word as far as I was concerned. The Drive was peopled by ardent and hardened typical Bath uppercrust devotees and my first partner was an ageing lady, possibly the wife of a retired Colonel.

"Are you Culbertson?" she asked me, to which I replied,

"No madam, my name's Waldron."

I had no idea as to the nature of my gaffe but soon realised that all was not well. Somehow I persuaded my partner that I

The author, with Civil Service colleagues, 1940.

was indicating a strong hand, we finished up having to make a Grand Slam and probably made no more than 3 Clubs! I could feel the shock waves spread before me as I made my way from table to table as the evening wore on bringing chaos and disaster wherever I sat. I enjoyed the whole episode but clearly my partners felt otherwise. Needless to say my two companions had planned the evening as a huge joke at the expense of the Bath residents.

Our wartime office accommodation was in Kingswood School, at the top of Lansdown Road, almost within sight of the Bath Racecourse. It was a series of rambling old stone buildings complete with School Chapel, far removed in every way from Great Westminster House, Horseferry Road, London. The enforced and hasty evacuation had inevitably separated husbands from wives and families, sons and daughters from parents and severed everybody from their established social roots. In these circumstances it was truly amazing how quickly we all seemed to settle down and as time wore on and accommodation was found for husbands to be re-united with their wives, life, apart from the black-out, assumed a degree of normality. Many of the wives were recruited as temporary staff as were many Bath residents who seemed only too pleased to supplement their retirement incomes. With the rapid expansion of the Navy the DNA had a vastly increased work load, temporary staff once recruited needed training and, as a consequence, I was promoted into the Executive Grade as an Assistant Accounts Officer with responsibilities to train and supervise a section of some dozen clerks, mainly temporary, into the arcane mysteries of auditing ships and dockyard accounts. The majority of my 'temporaries' were in the thirty and forty age group whilst I was a mere nineteen years of age, but despite this I can recall no resentment arising out of this age gap and I revelled in the responsibilities of my new position. The resultant pay rise was also much welcomed.

Other Admiralty Departments had also been evacuated to Bath, including the Department of Torpedoes and Mines in which my Uncle Ernest was playing an ever more important role in the countering of the German Magnetic Mine menace and, later as a co-developer of the Z Batteries, an anti-aircraft weapon using a cube of sixty-four rockets rather than the more directional and conventional shell from Ack-Ack guns. The Bath Meeting Room swelled in numbers as a consequence of the influx of Plymouth Brethren employed in the many departments of the Admiralty and much effort was expended in a vain attempt to 'keep tabs' on my activities. Frequent invitations to tea or supper as a means of keeping me within the 'circle' were as frequently declined, to the concern of the Brethren. My new found freedom 'at home' with my friends allied with the privacy of movement conferred on me by the black-out, enabled me to join the Little Theatre Club, a part professional Drama group, as an unpaid stage-hand. To this sanctuary I escaped of a Sunday evening after the Gospel preaching, sitting always near the door ready for a hasty exit. Evening meetings were rarely attended, the excuses being offered varying between pressure of work or the distance from my lodgings to the centre of Bath. I have no doubt that their worst suspicions were confirmed when a Leading Brother, the formidable and dour Mr Barnes – yet another Civil Servant – saw me one summer's evening wearing an LDV armband and carrying a Long Lee Metford rifle. I cannot recall the repercussions of this episode but suffice it to say that I refused to give up my membership and carried on as before.

Some explanation of these quasi-military activities is called for. The Local Defence Volunteers (LDV) were the forerunners of the Homeguard and the Admiralty had encouraged their staff to enrol in Sections set-up in each Department. This I had responded to and as a result we were equipped with ex-Boer War Long Lee Metford rifles, re-calibrated to use modern 303 ammunition. Thus armed we were taken to a naval rifle range for practice and those adjudged not to be too serious a menace to their fellow-men were allowed to retain their rifles as personal weapons. As a Section leader in the office I automatically became an LDV Section Leader in the Field and as such was supplied with a whistle and five rounds of ammo, one for each of our five-man squad. Came the day when a report was received of suspected German Paratroopers 'invading' the Downs near our offices. My section was summoned for action and off we tramped over the intervening countryside. My instructions were precise. On sighting the 'enemy' I was to pass one round of 303 to each man, rifles to be loaded, safety catches on and the enemy engaged. However no one was to fire until such time as I blew my whistle. Thankfully the observed parachutes belonged to an RAF crew who had bailed-out some distance away, contact with this potential 'enemy' was never made and we returned safely to base! The mind 'boggles' at the inevitable outcome had these indeed been German Paratroopers.

I was becoming increasingly dissatisfied with my role in the War and although not officially in a Reserved Occupation, arrangements had been agreed by 'the powers that be' to ensure that our services would be retained by the Admiralty. I had no wish to become involved in these behind-the-scenes arrangements and having already rejected pacifism as a valid response to the threat posed by Hitler, I was seeking an escape from this impasse. The events of June 1940, Dunkirk and Calais in particular injected a fresh impetus into my thinking. I felt a personal sense of loss when it became clear that my friend Cyril Pickering, serving with the Queen Victoria's Rifles had been captured at Calais and I regretted that when he and Jack Jonas had enrolled in the Terriers' I had been unable to follow suit. Throughout this period I visited Hilda at her family home in Pontypridd though not as frequently as I would have wished. My means of transport was by now a 150cc Coventry Eagle 2-stroke and to supplement my meagre petrol ration I ran this on long distance journeys on a mixture of paraffin, methylated spirit and naphthalene balls.. In order to start the engine on this mixture the sparking plug was removed, a squirt of lighter fuel injected into the cylinder, plug replaced and the procedure repeated should it fail to fire the first time round. It never failed on the second 'kick' and once warmed-up it ran like a dream. This took me to and fro both to Pontypridd and Portsmouth to see my parents but except for confiding my intentions to Hilda, I kept my own counsel both with my parents and my future 'in-laws'.

Two or three months elapsed during which time Hilda had come to Bath for a week's holiday and despite her father's opposition she stayed at my lodgings under the watchful eyes of Mrs Oborne who made it clear that no 'hanky-panky' would be allowed. My plan of action was clear. I would in due course volunteer for service in the Army which would inevitably result in my being thrown out of the Exclusives – in their parlance 'withdrawn from' – pacifism being one of the cornerstones of their beliefs. Hilda fully accepted this course of action fully aware that when the time came for us to be married she would likewise be removed from Fellowship – a prospect which she looked forward to with relief.

The final trigger for my volunteering for Army service came in the shape of an advertisement in the Motor Cycle magazine calling for volunteers from the motor cycling fraternity to join the Royal Corps of Signals as Despatch Riders. Armed with this advertisement I called at the Army Recruiting Office one lunchtime and a delighted Sergeant swore me in, gave me the King's shilling and wished me well and added that I should receive my 'papers' within two or three weeks. On returning to the office I reported my action to Head of Department, who was horrified and set about seeing if he could reverse my foolish action. Thank goodness such was not to be. I had taken the 'oath', received the King's shilling and that was that. The final act in this particular 'play' was yet to come and was in many respects the most emotionally demanding. My parents had to be told and this would without doubt cause them great pain particularly as my father had been a resolute and sincere Conscientious Objector in World War 1. The local Meeting Room also had to be informed of my action and this I realised would result in a harrowing arraignment before some of the Elder Brethren.

Looking back over nigh on sixty intervening years, I ask myself whether I realised at the time that this was to be the most significant and defining decision of my entire life? Shrouded in elapsed time such emotional experiences defy accurate recall and I must therefore guard against the temptation to attribute degrees of profundity and foresight beyond the workings of an immature mind. Nevertheless I was fully aware, to the point of physical nausea, of the shattering effect this decision of mine would have on my parents; aware also that the friends and relations with whom I had shared my life hitherto, would ostracise me and that my much loved Uncle Ernest would have to be faced as one of my 'judges' in the forthcoming confrontation with the Exclusive Brethren Establishment. Not least, faced as I was with the dubious distinction of being the only Exclusive Plymouth Brother to volunteer for H.M.Forces, was I right? What I did not forsee was the eventual destruction of my Christian faith based as it was on the inflexible foundation of Fundamentalism. But now to return to the realities of the situation as it unfolded in October 1940. I was a mere twenty years of age.

In what sequence the responses came I know not. My father wrote me a profound and incredibly understanding letter and the local Meeting summoned me to a Brother's house for the arraignment I had anticipated. The latter event I will describe first. Three 'Elders' were present of whom I can remember but two – Uncle Ernest and Mr Barnes. Three or four hours of readings, prayers and harangue ensued though with what purpose in mind is obscure as the 'dye had been cast'. My uncle a designer of torpedoes and mines was ,as far as he was concerned, a conscientious objector as, incidentally was his brilliant son who in due course played a leading role in the development of the British Atomic bomb. They saw no inconsistency in these attitudes. But I digress. Late into the evening with midnight approaching, I was finally warned of the evil of my ways and its inevitable consequences. Mentally and emotionally exhausted I rose from my chair to leave, at which point Mr Barnes stood-up, hands on lapels and in a sepulchral voice said 'Young Albert, this night thou must choose, God or Mammon'. I continued to the door of the room, opened it and looked back saying 'Its Mammon for me-goodnight'.

The front door was near at hand, I went out into the night to walk up a long front garden path. I was at this point near physical collapse, made doubly worse by the thought that God would strike me dead before I made my escape. I was at this stage of my life still a committed Christian but from that night onwards I re-examined, bit by bit, the foundations of my faith and found them wanting.

My father's letter reflected a totally different reaction to my chosen path. His reply, written in a copper-plate hand and in beautiful language derived from much bible reading, was in three paragraphs. The letter is sadly lost but its essentials for ever etched upon my mind. In his opening paragraph he firmly disagreed with my rejection of pacifism and supported this with copious quotations from Holy Writ. In the following paragraph he examined his own assumed failings in my upbringing, what he had endeavoured to teach me – more biblical quotations – where had he gone wrong. The final paragraph was nothing short of sublime and has lived with me and often guided me through the conflicts of my life. But my son, he wrote, I respect you, for what you have done you have done according to your own conscience. As I write these words I still marvel at the quantum leap that my father, steeped in pacifism, had to make in order to write those words. Throughout my life, particularly in industrial management situations, faced with seemingly intransigent attitudes, possibly from Trades Unions and drifting towards a condemnatory attitude, I have called for an adjournment. I have then asked my secretary for a cup of tea. My door has been closed and I would think of my dear 'old Dad' who confronted with a son whose alien views were at complete variance with his own sincerely held convictions – of a lifetime – could still write – 'I respect you'. What a remarkable act of tolerance and the associated paradox – it came from an Exclusive Plymouth Brother. Thank you father.

Little more needs to be added to this chapter in my life. My call-up to the Royal Corps of Signals duly arrived some three weeks later and I 'escaped' into a new life with many unanticipated challenges and disappointments. Hilda's parents, understandably, tried without success on this second occasion, to disentangle her from me, but to their eternal credit, throughout the War and even after our marriage, the door was never barred to me. For this I am sure I have to thank my loving and outgoing mother-in-law. I finally joined the Army on 25th November 1940 and on 27th November my father wrote me yet another understanding letter imploring me to 'keep short accounts with God'. On joining the Signals in London my records went astray and I was officially 'lost'.

By the time that this second letter arrived my father was dead – killed by an enemy bomb during a 'hit and run' raid on Portsmouth.

CHAPTER 6

Arch Plotter

Was there a sense of deja-vus as once again I boarded a train with a one-way ticket for London ? I doubt it entered my mind. The differences were many and not all were superficial. Some three and a half years previously I had travelled from Portsmouth to Waterloo on a Southern Railway electrified train; this time from Bath to Paddington on a 'real' steam train, no doubt pulled by a magnificent locomotive possibly of the 'Castle' class. Then I had been a puny boy of barely 5 feet in height and encumbered with a dreadful stutter; now a sturdier 5 feet 10½ inches almost freed from my speech impediment. Then a protesting spirit trapped within the caring but restrictive embraces of the 'Exclusives'; now a free spirit with the opportunity to shape my own selected approach to 'salvation'. Then a young lad travelling to the secure and ordered world of the Civil Service; now on a journey into the lowest ranks of the Army where all personal identity was to be speedily lost into the anonymity of my Army number – 2596687.

What did I expect on my arrival at the Old Brompton Road Underground Station where I had been instructed to report ? Maybe a kindly Sergeant sitting behind a reception desk who would greet me with a degree of warmth as befitted a motor-cyclist who had volunteered to help the Royal Corps of Signals in their 'hour of need'. It seemed obvious to me that a critical shortage of Despatch Riders was the raison-d'etre behind the advertisement in the Motor Cycle magazine which had finally precipitated my enrolment. If that was in my mind I was to be speedily disillusioned. The Underground Station was in fact the Control Centre for London's Air Defences. All the support services – telephones, communications with radar, gun and searchlight sites, plotting tables indicating the location of both friendly and enemy aircraft – were manned by 'Signals' personnel, whilst officers of the Royal Artillery directed guns and searchlights in accordance with the available information. In due course I had a fleeting look at this activity buried deep below London but before this and still dressed in 'civvies' I was directed to the 'Mess', just across the road, located in an ex- Joe Lyon's Tea Shop but regrettably without the 'Little Nippies' who were so much a feature of the pre-war London scene. I espied a quiet corner for my first Army meal only to be accosted by an irate sergeant. I had unwittingly invaded the sanctity of the Sergeants' Mess – not an auspicious start to my Service life!

Next day I was divested of my 'civvies' and kitted-out in an ill-fitting uniform as were other recruits who had appeared on the scene. Apparently the 'Signals', a supporting technical Corps, were at the end-of-the-line for new equipment resulting in a limited range of clothing and ancillary equipment. We were even paid an allowance for supplying our own boot and tooth brushes! That day we were whisked-off to Stanmore by truck and billeted in an empty, requisitioned house and set-to to scrub and polish it from top to bottom. No motorbike in sight. Unfortunately at this stage the Army 'lost' me – no Army number, no AB64, no pay – I was indeed a non-person for a short time and this came near to bringing about a personal tragedy.

Our induction into the demanding world of Army discipline continued under the stern gaze of sundry lance-corporals,

corporals and sergeants, whose sole purpose in life was, as far as we could judge, to make our lives as miserable as possible. No sooner was our house-cleaning fatigue completed than the 'boning' of boots, polishing of brasses and blancoing of webbing seemed to fill our lives, punctuated by marching up and down Stanmore Hill to Bentley Priory where Fighter Command provided Messing facilities for our Corps of Signals training Unit. Three times a day, usually in wretched December weather, we marched to the Priory for 'our daily bread', such marching being interlaced with 'Square bashing' en-route. A particularly 'sadistic' twist to these routines was applied should we perchance have infuriated our Drill NCO with an inept performance – an all too frequent occurrence. Then, when in sight of the Priory, relaxation and food, we would be 'about-turned', marched down the hill – shades of the good old Duke of York – and after further 'bashing', both verbal and drill, marched back up again to join the end of the queue for our midday meal. How we hated those NCO's.

Thus it was after an early morning's march from billet to Priory, the sun still not risen, we had barely 'drawn' our breakfast when I was summoned to the Corporals' Mess, an unheard of intrusion into our 'Private' time. What heinous crime had I committed ? Sadly to say the reason for the summons was no such trivial matter. A telegram had been received notifying me of the death of 'a member of my family'. In the short pause that ensued I immediately assumed that my mother had passed away, an eventuality I had been prepared for from an early age. Not so. My father had been killed in the course of a 'hit-and-run' raid on Portsmouth on the 5th December – by now it was some five or six days later – and was to be buried that very afternoon. My 'loss' by the Army had caused this delay and had also reduced me to a state of near penury. The human side of the much feared Corporals was now revealed. A quick 'whip-round' yielded more than sufficient pocket-money for my day's travel, a Railway Warrant, Waterloo to Portsmouth return, had already been prepared and a 15cwt truck was standing-by ready to rush me to Waterloo Station to catch a train the time of which they had ascertained. I was urged to finish my breakfast – wise advice followed with much difficulty- and then away to the billet to collect personal kit and on to Waterloo. Of the subsequent journey in a physical sense, I recall absolutely nothing, but the seemingly interminable journey was thought-wise taken-up with recalling 'the last time I had seen Dad'. Is it not always thus in such circumstances ?

My mind took me back to my last visit to home from Bath some weeks before I joined the Army. I was still, nominally at least, a Plymouth Brother and I gave my parents no inkling of the distressing decision which I was to take within a week or so. I attended the Sunday Gospel Service in the Cosham Meeting Room at which my own father was the preacher. His preachings were always well attended by the young people from other Meetings in Portsmouth as he was renowned for his directness and brevity. His own conviction was that if anyone 'preached the Gospel' for more than twenty minutes, that preacher was getting between God and the listener. On this particular evening father took as his text the Epistle to Jude, verses 20 – 25 with especial emphasis on the words, "Now unto him that is able to keep you from falling..." He finished speaking and after a short prayer announced the last hymn, "All hail the power of Jesu's name'. He always started his own hymns and as he drew breath I sensed that he would start the well-known tune -'Miles Lane'. Before his first note could ring-out I preempted him with that rousing tune – surely composed specifically for that hymn – Diadem, with its rolling repeats of the last line of each verse. His face lit-up with pleasure and at the end of the last verse he sang a repeat then leapt from the platform, rushed down the aisle, hugged me for starting a tune he loved but had never 'had the nerve' to start. Such was to be my final and most vivid memory of my father.

Of the physical aspects of the journey to the front-door of our house in Cosham I remember nothing. Suffice it to say that I arrived at exactly the same moment as the hearse, which was kept waiting for a few minutes for me to ready myself for the journey to the cemetery some three miles away. In those few minutes I was told by my sister that my father having been slandered and vilified by fellow Brethren had been in serious dispute with them at the time of his death and warned me that this could have an impact on the forthcoming Service. In the event ,with one minor exception, all passed-off with a moving and sincere tribute to my father made the more poignant by a large attendance of coal-heavers from the Goods Yard where he had been the Railway Agent and was respected for his fairness and scrupulous honesty. Some of these rough and ready men were in their Sunday-best but others straight from their coal drays were in their blackened working clothes. A sincere farewell to a truly fine man.

I suppose I had been granted 48 hours Leave but again of my return to Stanmore I remember nothing. The forced detachment of Army life no doubt dulled the pain of my personal loss, training continued, a cold and miserable Christmas ensued and early in the New Year I was posted to the Signals Unit manning the Bristol Control Centre for Anti-Aircraft Defences. This turned out to be but a brief, temporary posting punctuated by only two notable experiences – very much a plus and a minus. Despite my earlier avowal to 'serve Mammon', I was still a committed Christian and one Sunday evening found my way into a near-by Chapel, possibly Congregational. It was all but full and I found myself a seat near the door. There was an obvious delay in starting the evening Service, due, as was explained by the Minister, to the non-arrival of the organist. Could someone in the congregation officiate ? There was no response, the American organ beckoned, and after a long pause I stood up and clumped my way up the aisle in my heavy Army boots. The Service proceeded without a hitch, I was back in a world far removed from the Army, was thanked profusely, taken out to supper and asked if I could stand-by for next week. 'Twas not to be. I contracted scabies from dirty blankets and then endured ten days of unmitigated misery in a Medical Centre – not a hospital – anointed in sulphur ointment, walking across a snow covered yard to ice-cold showers and generally treated as having committed a criminal act. The sense of isolation was further compounded by the absence of incoming letters and, although we were happily unaware of this at the time, none of our outgoing mail was ever posted and once again I was 'lost' for this period much to my widowed mother's distress.

Yet again I was on the move. No sooner had I been released from medical attention than I was on my way to Plymouth for a 'permanent' posting to a Signals Company engaged in the Ack-Ack defences of Plymouth. For some four or five weeks I underwent a purely Signals training – morse code both by key and lamp, some rudimentary instruction in 'line' work which, much to my joy necessitated swarming up and down telegraph poles, basic map-reading, co-ordinates, Eastings and Northings and other essential basics of telecommunications. It was an ex-Territorial Unit consisting mainly of erstwhile Post Office employees and was as helpful and pleasant a bunch of men as you could wish to meet.

This training completed I was transferred by truck to Redruth in Cornwall, passing Bodmin Moor covered in snow in mid-February. The Brigade Headquarters of Ack-Ack Command Cornwall was located in a spacious house near the centre of the town and here I was soon established, in company with an experienced operator on a double-cord switchboard. Hardly a motor cycle and not particularly suited to an operator whose stammer returned when under pressure!

I was at last introduced to a despatch rider's bike, a Panther 600cc 'sloper', far removed from the 150cc 2-stroke that I had been wont to ride some three months previously. Not surprisingly, lacking any instruction, the test was not, repeat not, a success and I was returned to switchboard duties. How to escape was uppermost in my mind. Protests would, I knew, be futile. Fate, in the shape of the two most senior officers, the Brigade Major and his Brigadier, played into my hands.

The former, Major Rochier, was urbane, suave and obviously well-educated. By complete contrast his senior officer, Brigadier Barbarry, was crude, ignorant and virtually without social graces. Likewise he had no consideration for the many difficulties involved in the running of the Brigade, particularly those in the field of communications where we were entirely dependent on GPO lines. The day came when, frustrated by his demands for an instant line to his wife, Emma, at a suitable juncture when both the 'Brig' and his 'Major' were awaiting long distance calls, I deliberately cross-plugged them thereby creating the illusion that each was taking their long-awaited call whereas in point of fact they were talking to one another across the passage-way. The exchange lasted but a few moments, was hilariously funny whilst it lasted – a clash of personalities – the outcome predictable in that I was physically removed from the switchboard by an enraged Brigadier and declared 'persona non grata'. Yet another posting was forthcoming, this time to a detachment at Feock near Truro where a small Control Room had been established for the Ack-Ack defences of Falmouth, but before leaving Redruth one more episode in my 'personal journey' needs to be recounted.

From my departure from Bath at the end of November until my arrival at Redruth in mid-February, the Brethren had followed my movements and under the guise of offering hospitality, had pressurised me into occasional attendances at Meeting Rooms, admonished me on my departure from 'the paths of righteousness' and even suggested that I back-tracked and applied to be considered as a Conscientious Objector! When at Redruth two of the local Brethren came into my billet and once again the familiar pattern was to be repeated I finally had had more than enough. I threatened to report them to the Army authorities for entering a Service billet without permission and with the intent of subverting a soldier in the course of his duties! Never again was I approached by the Brethren nor in the intervening 58 years have I set foot in one of their Meeting Rooms. In 1942, my wife, then in the ATS was forced to cope with very similar experiences.

Feock in the early Spring had much to commend it. The Control Room was the ex-Drawing Room of the Dower House on the Trelissick Estate, King Harry Ferry was just down the road and the woods were about to become carpeted with thousands of wild daffodils. We were housed in a small group of Nissen huts set in a small woodland adjacent to the Dower House. One of our number, a married man, had rented an old fisherman's cottage on a quiet sea inlet, was granted a 'sleeping-out' pass and with the co-operation of his long-suffering wife invited two or three of us for evenings 'away from it all'. What a blessed relief. However the Control Room was no dungeon, but enjoyed wonderful views through large French windows onto the somewhat neglected gardens, and was equipped with the usual raised dais for the controlling Royal Artillery officers, commanded by a Captain Adamson who wore RFC wings on his tunic as an ex-pilot of World War 1. Some months later he was to have a considerable effect on my Army life. In addition to the dais was a multi-phone board linking all gun, searchlight and barrage balloon sites to 'Control', a telephone switchboard for outside communication via the GPO and finally the twelve foot or so square map table, suitably inclined towards the dais for ease of reading, setting-out the coast line of our Sector, the sea beyond and some of the French coast.

Angled wooden blocks some three inches in height and having three grooves, sat on adjacent racks awaiting an enemy raid when the relevant information – raid identification number, number of aircraft and estimated height – would be inserted into the appropriate grooves and the block located on the map table at the correct grid reference. Two 'plotters', one left and the other righthand side of the table, equipped with headsets, received this information from RAF Rudloe, the radar centre, and then set it up and up-dated it as required. Apart from occasional relief duties on the multi and switchboard equipment, I was trained to become one of the regular plotters and was fortunate to serve my 'apprenticeship' in such a friendly and relatively quiet environment as the Falmouth area.

I doubt we fully appreciated the delights of the Cornish countryside and seascapes. Truro was not easily accessible, the nearest bus stop being some one and a half miles distant, though a pub in a nearby village provided some facilities for relaxation. A side-line was provided by one of our Detachment who in civvy-street had been a professional flower packer. How the packaging materials were obtained is a mystery but throughout the early Spring boxes of daffodils were posted to destinations throughout Britain, doubtlessly to the delight of families at home. Sadly for me this peaceful and rural bliss was about to be rudely shattered. Air attacks on Plymouth were anticipated and in the late Spring I was returned from Detachment to H.Q. located in an old Redoubt in Crownhill, Plymouth, itself a relic of Napoleonic days and his threatened invasion of England.

Looking back I now realise that the exchange of personnel between Plymouth and Feock was a wise policy in that Feock was used both as a training detachment away from the pressures of Plymouth and, training once completed, as a rest and recuperation posting following a spell in the intense pressures of the Plymouth Control Room located deep within the bowels of Fort Egg Buckland, yet another relic of Napoleonic days defences. I returned to H.Q. without any evidence that my training had borne fruit. I consistently and deliberately failed all Trade tests, had no Proficiency badges and had no ambitions to rise above the basic rank of Signalman. I had no financial incentives as I was in receipt of 'Balance of Civil pay' which gave me an income the equivalent of a well-qualified Sergeant. Nevertheless I found myself pitched into the role of the number 1 plotter at a time when the second wave of blitzes was about to engulf Plymouth. It was both physically and mentally demanding. Night shifts were of fifteen hours duration, acceptable if a quick 'kip' could be taken during a quiet spell, but during a full-scale air attack this was obviously impossible. The sheer density of the attacking aircraft and the consequent number of 'blocks' on the map table, created a situation in which the flow of information from Rudloe could not possibly be absorbed and actioned in its entirety yet it was essential not to lose control. Somehow one learnt to focus on the most vital information and to discard other data. On occasions it was necessary to supplement plotted data with shouted information to the RA officers on the dais but despite the pressures, at no time did I experience any sense of panic. In short it was a wonderful team effort where rank counted for little and individual commitment was all-important. Unbeknown to me it must have been a character building experience! After a night of intense activity it was march back to camp, breakfast and then the regulation eight hours asleep but for me and my bosom pal Chris Mallinson, following a quiet shift, we had other less orthodox ideas.

Crownhill Redoubt, our Nissen hutted camp, was a moated fort with a fixed planked bridge guarded by sentries from an Infantry Battalion who at the same time checked our movements in and out. The moat though deep with steep walls was at least dry, and Chris and I decided that an emergency exit and entrance was essential for our well-being. Fortunately there was one significant break in the outer wall where the brickwork had collapsed so if access could be gained to the moat, escape was feasible. With the aid of obsolete cable a 'rope' ladder was constructed, hidden in bushes above the moat and by this means on suitable occasions we unofficially left camp for hiking in the nearby countryside. Pre-walk planning was essential. Food for the day was purloined from the Quarter-Master Sergeant's private and illegal store when we volunteered for Sergeants' Mess fatigues. Evading the duty sergeant on our 'sleeping' days was more tricky. Following breakfast, into bed fully clothed except for boots, sergeant's check consisted merely of counting 'heads', that once completed a dash for the bushes and the ladder. During that Spring and early Summer we wore out many a sole of our Army boots to the mystification of the CQMS, visited distant places thanks in the main to friendly lorry drivers and never failed to return on time. Bread and ale in the refectory at Buckfast Abbey and meeting the parents of my Civil Service friend – Jack Jonas – are stories in themselves but not for the telling at this time. One episode merits inclusion as it epitomises the ingenuity and courage of victims of the blitzes. Returning across the Moor in the gathering darkness one Summer's evening, in the vicinity of Dewerstone Rock Chris and I were intrigued by what appeared to be a blazing fire some distance away, apparently in the middle of 'nowhere'. We diverted to investigate and as we drew closer a group of people could be seen, seated around the fire within a circle of tall vans 'corralled' on an outer perimeter. We entered the 'magic circle' and as our Ack-Ack shoulder flashes became illuminated by the leaping flames we were surrounded by cheering citizens of Plymouth. Bombed -out from their homes they had equipped furniture vans with bunk beds and each evening left the city for the peace and safety of the Moor. Royalty could not have been more generously entertained than Chris and I on that memorable evening – never to be forgotten.

Periodic 7 days leaves enabled me to keep in touch with my mother in Portsmouth and Hilda, now back home in Pontypridd and employed as a wages clerk in a near-by Power Station. I could well describe these as 'busman's holidays' as on each occasion my spell of leave coincided with a blitz – Portsmouth and Cardiff, the latter being ablaze as I left the Station and set-off on the 13 miles to 'Ponty'. By a combination of walking and 'thumbing' I arrived at 16 The Common, Pontypridd where the family were out on their balcony watching the lurid sky as it hung over distant Cardiff. At about this time my mother found a suitable tenant for her house and moved to Reading, sharing accommodation with my sister who was teaching at Goring-on-Thames and remaining there for the duration of the war. Meanwhile my brother had left the Police having volunteered for RAF Aircrew duties. To his disappointment he was trained as a Navigator and was in the fullness of time posted to a Glider Towing squadron as a Flight Lieutenant.

I was by now 21 years of age and beginning to find Army life irksome and unrewarding, though ameliorated by my unofficial adventures with Chris. These came to an abrupt end as I was once again posted to Feock and resumed my duties in the Dower House Control Room. Nothing much had changed in my absence except that a Signals Lieutenant had been installed as i/c of our Detachment. I was not to know that he was to become a major influence on my future though entirely by 'default'. Life was somewhat enlivened by duties as a temporary 'linesman' under the direction of a jovial Sergeant and occasional forays to repair broken land lines provided the opportunity to enjoy the pleasures of Falmouth beach. Captain Adamson was still the senior Royal Artillery officer and as such in charge of the Control Room. He was 'supported' by several Subalterns of somewhat ancient vintage, many of whom were destined to be 'weeded-out' when the Churchillian purge took place. One of these 'ancients' was particularly incompetent and it was obvious to the Signals' Detachment that he was shielded from potential crises and rarely left on his own. A stratagem to 'smoke him out' was devised in which I was to play the key role.

Adjacent to the Control Room was a lounge where the RA officers relaxed unless the situation as set-out on the map table warranted their presence on the dais. The Corporal i/c the Signals' shift would ring a warning bell should an enemy raid be indicated. On the day in question, when taking-over for the morning shift, it was ascertained that 'papa' as we called him, was alone on duty and Operation Lone Daylight Raider was duly launched. As senior plotter it fell to me to inaugurate a 'raid' over the French coast, one enemy aircraft, height 1000 feet, direction northerly on a track heading for the Falmouth area. With headset on as if listening to Rudloe I moved the imaginary raid closer and closer to the point where 'papa' had to be summoned to take control. Panic ensued. Guns were not in fact alerted though the multi-phone operator simulated such action. The 'enemy' aircraft was now on track to pass over Feock but before the prank could get out of control I announced that the aircraft had flashed IFF, thereby being identified as 'friendly' – raid cancelled. 'Papa' was all but a nervous wreck and repaired to the lounge doubtless for a stiff Scotch and Soda. Operation successful!

Retribution was to follow shortly. Next morning I was summoned to appear before Captain Adamson who with the Control Logbook by his side 'suggested' to me that the so-called 'raid' had been a contrived affair. Without hesitation I agreed and admitted to my key role in the whole affair. He commented that it was his responsibility to judge the abilities of his officers and that any repeat of this episode would not be tolerated. The mildness of his reprimand surprised me, particularly as I was aware that my actions in all probability warranted a court-martial. To my further surprise he asked me to sit down as he had some personal advice to proffer. He had come to the conclusion that I was not using my abilities to the full and I should seek a more challenging role within the Services. He then drew my attention to a War Office Order calling for volunteers to transfer to the RAF as aircrew and suggested that I might well apply. He would do all in his power to facilitate an application and would loan me an instruction manual on 'How to Fly'. His previous RFC experiences and his personal enthusiasm for flying had clearly motivated him into giving this personal and friendly advice, and I had no hesitation in making the necessary application for transfer.

In due course I was instructed to travel to Oxford to an Aircrew Selection Board, the immediate hurdle to be surmounted en-route to becoming a Spitfire pilot! Such were the dreams of young men in those post-Battle of Britain days. At Oxford, the written tests, basic Maths, English and a smattering of General Knowledge presented no problems and to my relief I likewise had no difficulty in being passed as 'Medically fit for aircrew'. The final interview with a Board of senior RAF officers awaited but as a 'W' for Waldron I was, as ever, at the back of the alphabetically organised queue. The RAF corporal organising the waiting 'hopefuls' confided to me just as I reached the door that 'them in there' were all in a foul mood and no one had been passed that day. Armed with this negative information (almost certainly spurious) I marched in, saluted and was instructed to sit down. The Presiding Officer, an Air Commodore, identified me from his papers as Signaller Waldron. I remonstrated pointing out that I was in fact Signalman Waldron at which point an argument ensued and I had to explain the subtle but vital differences between a Signaller and a Signalman – the former an infantryman trained in signal communications and the latter a member of 'His Majesty's Royal Corps of Signals'. Point made I settled down to a barrage of questions from the five members of the Board until I ventured to suggest that there should be a reversal of roles and as it was me who was volunteering to place my head on 'the chopping block' might I ask a few questions. Clearly surprised, they assented, I hastily invented a few relevant questions and the interview was concluded. I had passed!

I returned yet again to Feock, celebrated my success and imminent departure with a party in a Truro pub and awaited with a high degree of impatience the necessary 'papers' transferring me to the RAF for aircrew training. Captain Adamson took me aside and congratulated me on my success and I thanked him for his help. Some weeks elapsed and in late November a telex from HQ Plymouth was received. A direct quote is essential. "Signalman A.Waldron, 2596687. Transfer to RAF cancelled by Order of War Office. If he still wishes to fly arrangements will be made for his transfer for training as an Army Driver Pilot. Reply immediately." What is a Driver Pilot? I asked of my Signals Lieutenant ? I was reminded that the Germans flew troops around in Junkers 52/3's. The British Army was setting-up a similar Airborne Force and to distinguish the pilots from fighter and bomber pilots the term 'Driver Pilot' was being adopted. This sounded feasible and any form of flying seemed preferable to 'Arch-plotting', so though disappointed I volunteered once again, this time for an aeroplane, not a motor-cycle Within a few days my papers came through transferring me to 16 EFTS, Burnaston, Derby and it required but a minimum of intelligence to translate EFTS as Elementary

Flying Training School. The auguries were favourable and in January 1942 I travelled to Plymouth for the last time, was 'cleared' through HQ and put on the train for Derby.

In contrast to G.B.Shaw's 'Chocolate Cream Soldier' I was the 'Bottled Bass Soldier', carrying a tumbler in one 'pouch basic' and a bottle of Bass in the other. By the time Birmingham was reached my reservoir of Bass was long-since exhausted and I enquired of a porter if I had time to 'hop across' the bridge and get a drink. A long wait was indicated so thus encouraged I crossed to the Station buffet for replenishment only to see the train depart taking with it all my kit. Shattered, I went into the RTO's (Rail Transport Officer) Office, got a bollicking for not wearing a forage cap, explained that it, plus all my kit, was en-route to Derby and was instructed to catch the next local train. I duly arrived at my destination, had a further dressing-down by the Derby RTO and was then directed to the YMCA just outside the Station to await transport to Burnaston. I all but boarded an Army truck full of prospective parachutists – not for me under any circumstances – returned to the YMCA and met a fellow Signalman though from another Unit, who was also bound for Burnaston. On exchanging 'notes' ,it transpired that for 'Driver pilot' read 'Glider pilot', an entirely different story and one which I would have rejected out-of-hand. Flight without an engine was a form of idiocy in which I had no intention of becoming involved. Indeed, had I not been bereft of all kit, not even a forage cap, razor or toothbrush, I might well have about-turned there and then. Instead I boarded the appropriate truck and Burnaston was reached on a bitterly cold day, the 18th January 1942. My subsequent training as a pilot will fill yet another chapter in this saga.

CHAPTER 7

Rough Hew Them How We Will

The 'divinity that shapes our ends' had certainly seemed to frown on me as I arrived at Derby on that cold and bleak January evening in 1942. My kit was lost and the unwelcome prospect of flying an Army glider loomed large on the horizon. The 'frown', if it had ever been, was however short-lived. My kit, found at Preston, was re-united with me within 24 hours and in that time the immediate benefits of 16 EFTS, Burnaston far outweighed the distant prospect of glider flying. After all that 'bridge' could be crossed in three months time after I had qualified as a Mile's Magister pilot, an objective which I confidently expected to achieve.

At this time the concept of glider borne troops was shrouded in secrecy and, as a consequence, details of our training programme were somewhat vague. Certainly 3 months were to be spent learning to fly a light aircraft, a Miles Magister. If successful, we were then destined to spend 4 weeks at a Glider Training School (GTS), followed by 7 weeks at a Glider Operational Training Unit (GGOTU) flying an Army Hotspur Glider, designed to carry 8 troops. At the end of this time we would be awarded our 'Wings'. Beyond that our future was shrouded in mystery. But to return to the present.

We were billeted in one of the 'Houses' of the erstwhile Repton College, with the luxury of double bedrooms one of which I shared with Roddy Stock, one of the real characters of Number 1 Course which had preceded us by two weeks. Both of our Courses consisted of some 24 volunteers posted directly from a diverse range of Army Units – Infantry, Artillery, Signals, Engineers et al. All subsequent volunteers were 'doomed' to experience the rigours of a Guards' type Depot at Tilshead, a desolate Camp uncomfortably close to an artillery firing range on Salisbury Plain. So close in fact that on the 8th January 1943, three corporals were killed by a 'short shell' which landed in the Medical section. Such a Depot regime I would not have survived and the humiliation of RTU (Returned to Unit) would have been the inevitable consequence. 'Fortune' had indeed favoured me.

On arrival at Burnaston we had all been promoted to the rank of corporal and received Flying Pay – wealth indeed. The 'old boys' who had 'done for' our schoolboy predecessors now appeared and offered their services for a modest wage. In no time Roddy and I had trained one such individual in the duties of a batman, an unheard of acquisition for a mere corporal and contrary to Army Regulations. With no little ingenuity we managed to retain our 'Jeeves' despite the unwanted interference of one bloody-minded officer!

The aerodrome itself, a mile or more away, was pre-war, a private airfield owned by one Roy Harben, who had been granted the rank of Wing-Commander but left to run the Flying Training School on a commercial basis. With the exception of the RAF Instructors, all other personnel were civilian – canteen, aircraft maintenance, ground subject lecturers – and this unusual organisational mix created a unique atmosphere in what was basically a Service situation. When flying we all 'clocked' on and off, punching our time-cards as if we were factory workers. Roy Harben was then apparently reimbursed by the Air Ministry at so-much a flying hour. Small wonder that shortly after we arrived and the grass airfield was buried in a heavy series of snowfalls, we were set to work to shovel a landing strip clear so that flying could proceed. So keen were we to get into the air that this chore was undertaken with enthusiasm and cheerfulness and on the 31st January I had my first taste of flying with my Instructor Pilot Officer Woolley. Once again I was lucky, as a more dedicated and patient Instructor could not be imagined and thanks to his skills and encouragement I went 'Solo' on 18th February after nine hours and fifteen minutes flying time. Not an outstanding performance but better than 'average'.

My first solo, in a Miles Magister or 'Maggie' flight trainer.

At this juncture something needs to be said about the 'Maggie' and its flying characteristics. It was designed and produced as a Primary Trainer for the RAF but unfortunately on Test it proved to be 'stable in a spin'. This may sound innocuous but in fact is potentially lethal. Once in a spin none of the controls would extricate it, it then went into a 'flat spin' from which it would not emerge before hitting the ground. When tested, as the pilot prepared to bale-out, the aircraft came out of the spin due to the turbulence created. This led to the addition of lateral fins on the upper side and length of the fuselage, the turbulence thus created in a spin prevented the onset of a flat spin and in such situations straight and level control could be regained. It was still somewhat lively in a spin and during our training much attention was paid to Exercise 10 – spinning left and right. Nevertheless the first fatality in the Regiment occurred at Burnaston when on 25th March 1942, Corporal Chaddaway, a member of Number 1 Course, spun-in near Derby and was killed. This episode shook all of us 'to the core' – at last we realised that life was not all fun – and Wing Commander Harben took it upon himself to give us all a tough talking-to, berating the dead man for having destroyed one of his beloved Maggies and telling us to get away for the week-end and come back in a positive frame of mind.

I had two close friends – Geoff Nelson, subsequently killed at Sicily and Bill Denholm who suffered a similar fate on the Rhine crossing. Bill and I, taking the Wing-Co's advice, set-off for Derby seeking hotel accommodation for the week-end. There was none to be found even in the Royal Standard Hotel which turned out to be merely a pub. By this time we had imbibed a few pints and as we sat with our drinks contemplating a return to Burnaston we looked up at the glass case above the bar, full of stuffed pike, some of which were veritable monsters and one of which appeared to move a fin. Obviously more under the influence than we thought, we commented to the landlord, a genial and portly man, who laughing, invited us in for the next Friday evening promptly at 8 o'clock when all the pike in the TANK would be fed with small fry which he would have caught that day! At this point the landlady appeared, as genial as her husband and asked if we were the young men enquiring for a room. A sad story emerged. Both of her sons had been lost in tragic circumstances in early childhood, their bedroom was available and would we please be her guests.

From then on until the end of our course we stayed at the Royal Standard, plied with the most wonderful food and an excess of drink, never allowed to pay a penny and she wept when we completed our training and left for Oxfordshire. To our shame we failed to maintain contact.

But to return to life at Burnaston and Repton. In due course the 3rd Intake arrived, somewhat delayed by their enforced sojourn at Tilshead, blancoed up to the eyebrows, boots gleaming and, all in all, looking more like recruits for the Guards than potential pilots. We were appalled at this apparition which bid fair to destroy the camaraderie and free spirit that had developed within the first two Intakes and had flourished in the informal ambience of Burnaston. Worse was to follow. Our senior Course officers, Major Willoughby and Captain Viney, had held us on the lightest of reins consistent with an acceptable level of discipline, reasoning no doubt, that we needed all our powers of concentration to master the art of flying whilst at the same time absorbing the mysteries of Meteorology, Airframes,

---

Opposite: Number 1 intake, Glider Pilot Regiment, Brize Norton, 1942.

DENHOLM
GUTHRIE
PEARSON
TAYLOR
WALDRON
CRANMER
DOIG
NELSON
ROBSON
TORRANCE
WHITE
No. 1 INTAKE

Airmanship, Theory of Flight and Morse code. Now, with the arrival of the 3rd Course a young blonde Lieutenant imbued with the Guards' ethos of Tilshead was let loose upon us, presumably to shake us out of our sloth and impose a blanco and bull-shit regime. He was soon to be nicknamed 'Goering'.

Number 2 Course seemed to bear the full brunt of this officer's crusading zeal and clearly something had to be done. A meeting was arranged in our local pub – was it the Shakespeare – and the 'Bulshie Baiters' came into being with but one term of reference – defeat 'Goering'. The many machinations which we devised to disrupt 'superfluous' parades were both amusing and to a large extent successful, but the moment of our ultimate triumph, unexpected and hilarious, needs to be placed on record. The setting for this episode was an instruction to parade for 'Communications Drill' which consisted of drilling a Squad at long-range presumably with the intention of developing the stentorian voice beloved of Sergeant-Majors. A more dreary and pointless exercise it would be hard to imagine and totally irrelevant in the context of 'Pilot training'. Three factors contributed to our success. The location of the drill over which we had no control, the selection of the 'left-marker' and the 'volunteer' for the drilling duties for which we had contingency plans. I was the 'marker' and my great friend – he of the choirboy's voice – Geoff Nelson filled the role for 'trainee Sergeant-Major! It was agreed that under no circumstances would we ever hear his words of command. 'Goering' had demonstrated how this training would be conducted and Geoff duly took-over. Left-turn, quick-march was given at short range and obeyed and as we marched off the boundary of the actual flying field loomed in the distance.

"About-turn!" shouted Geoff: no response.

"Louder!" commanded 'Goering': no response.

Like the 'Charge of the Light Brigade", Number 2 Course marched on, not 'Into the Valley of Death' but across the Flying-Field into the path of Maggies taking-off and landing!! Flying Control could scarce believe their eyes, red Very lights were fired to stop all flying and by this time Geoff had been sent running after us, we duly 'about-turned' and marched back to our starting point. At this point an irate Wing-Commander Harben arrived on the scene. As an Honorary Officer he was clearly uninhibited by accepted Service etiquette, dressed 'Goering' down in no uncertain manner in front of the Squad. That marked our last parade save for an inspection by the Commanding Officer, Glider Pilot Regiment, Lieutenant-Colonel J.F.Rock whose untimely death in a flying accident at Shrewton on 27th September 1942 deprived the Regiment of their finest officer.

Meanwhile, despite these petty attempts to remind us that we were soldiers, our main energies were concentrated on the main objectives, to succeed as pilots and pass the forthcoming Ground Subjects examinations. My instructor, as I learnt later, had high hopes of me carrying off the prize for best pupil but sadly I let him down, On a cross-country flight on a somewhat murky day I got lost, came perilously close to flying through a balloon barrage and finally forced-landed near a farm. The actual landing entailed side-slipping to avoid high trees and touching-down in a small field and I was pretty 'chuffed' to make a perfect landing. An Air Training Cadet appeared as if by magic, gave me a map reference and a bearing to get be back to base – a clever lad. I then committed the unpardonable 'crime' by attempting to take-off again but even worse, I tried to do this down wind. I crashed into the hedge though with no injuries to self and not much damage to the Maggie which was subsequently flown out of the field by my Instructor. I made a futile attempt to lie my way out of the situation which could well have seen me 'Returned to Unit'. I survived, thanks to the intervention of PO Woolley, but although coming top in Ground subjects my misdemeanour denied me the award for 'best pupil as my rating as a pilot had been adversely affected. I passed-out as an 'Above Average Pilot' and this included an 'Exceptional' rating for aerobatics, a quality quite unsuited for glider flying! Unbeknown to me at the time efforts were made to secure my release from the Army for training as a fighter pilot, but to no avail. On 22nd April 1942 I left Burnaston with the firm resolve that glider flying would play no part in my future and when on the 28th of that month I arrived at Number 1 Glider Training School, Thame near Oxford and became acquainted with the Hotspur Mark 2 Glider, a somewhat basic aircraft, that resolve gained added determination.

Our arrival at Thame betokened a return to 'normal' Service life. No more luxury billets, 'batmen', or a civilian owned aerodrome, instead a hutted camp on the 'drome, all Service personnel, Army and Royal Air Force, certainly no unofficial batmen and, from time to time parades and early morning PT under the eye of an RAF Sergeant. Somehow we persuaded him that sitting in a hut at 6 o'clock in the morning drinking tea was much more civilised than physical 'jerks' in the cold morning air – the Bulshie Baiters had not lost their touch! The 'drome itself was small at this time – to be extended in the near future – and was bounded on one side by a railway cutting over which, depending on the wind direction, the approach and landing had to be executed. My first impression was of a

The Hotspur.

monster with its mouth wide open waiting to devour the unwary victim.

The Hotspur glider to whose peculiarities we had to adapt, was a sleek, wooden aircraft, with a wing span of 45'10" and a length of 39'3". It was built to carry 8 soldiers in cramped conditions and when fully loaded weighed some 1.6 tons. Its flying characteristics had nothing whatsoever in common with those of our late 'friend' the 'Maggie' in whose company we had learnt lightness of touch, sensitivity and respect for its skittish ways. By contrast, the Hotspur's flying controls – throttle not required – were crude, insensitive and basic in design. The joystick was a short broom handle which certainly did not respond to finger-tip control; the rudder controlled by metal slides was best manipulated with standard issue Army boots to overcome its tendency to stick in some off-set position, whilst the flaps were raised and lowered by a simple cable operated by a double wooden lever arrangement by the pilot's right hand. The 'Tugs' were a mixture of Hawker Hectors, Hinds and Audaxes which gave an impression of a World War 1 aerodrome. In fact the Hector in particular was an excellent towing aircraft with an abundance of power and a rear cockpit in which we 'thumbed' trips whenever possible. The general scene having been set it only remains to describe how 'fate' once again intervened to frustrate my resolve to leave Thame at the earliest possible moment and certainly before I was exposed to the trauma of a solo flight in this basic and crude 'motorless aircraft' as my logbook describes it.

A mere two days after my arrival having met my unsuspecting Instructor, Pilot Officer Hillman, I was seated in a glider behind him – instructor and pupil sat in tandem – for my first trip designated as 'air experience and familiarization with controls'. I was anything but impressed with the short 10 minutes trip and frankly scared when we approached over the 'yawning' railway -cutting, the wind being in a perverse direction. Five more 10 minute trips followed in quick succession with me reluctantly playing an ever-increasing part in the actual flying and as the glider was positioned for yet another trip and the tow-rope attached, exit my Instructor to be replaced with a sand-bag as essential ballast. Clearly a solo trip was in the offing and the time for protest had arrived. It fell on deaf ears. It transpired that PO Hillman, son of the Hillman of the Airways Company of that name, was anxiously waiting to return to 'real' flying, I was his last pupil and like it or not I was going solo there and then. There is no greater motivator than self-preservation, I was wholly at the mercy of the 'tug' as regards take-off, once airborne a safe landing was uppermost in my mind, cast-off, downwind, crosswind, turn into wind, yawning railway-cutting dead ahead, not too high, not too low, bumpy but safe landing, delighted Instructor and there I was 'hooked'.

Four weeks later on 27th May 1942, with 12 hours and 15 minutes of Glider flying under my belt, in company with the survivors of Number 2 Course – one or two had been RTU'd – I was off to No 102 Glider Operational Training Unit (GOTU) at Kidlington, where the next day but one we started our Operational training. This was to be a seven weeks course with a more variable and demanding training schedule culminating in the much coveted award of Glider Pilot Wings for the successful pupils on the 8th July, to be followed by 'end of the course' leave. Meanwhile my fiancée had been 'called-up', had joined the ATS and was undergoing her own training course at Nottingham where she took top honours. Needless

Hawker Hind (above) and Hector (below)

to say it was by now my firm intention to secure my Wings, all thoughts of 'ducking-out' now forgotten, so with no thought of failure in my mind we planned our forthcoming marriage for 11th July, a mere three days after the Wings' Parade.

Our previous Hawker bi-plane 'tugs' were now replaced with the Lysander, a high wing monoplane with an incredibly low stalling speed, so low in fact that for sheer devilment our tug pilots – mainly on 'rest' from other duties – would stall us off-tow, though always within gliding distance of base. A few Tiger Moths were provided mainly for relaxation and pleasure and although they never replaced the 'Maggie' in my affections, it was sheer joy to take to the air on a fine June evening with the added luxury of an engine after a day of arduous Hotspur training. The training schedule differed widely from that at GTS, and was clearly experimental to some degree.

The point of attachment for the tow rope was vertically beneath the wing at the base of the fuselage – known as belly-tow – and although deemed to be theoretically correct it had an unfortunate 'kite' effect and as towing speed increased it became more and more difficult to keep the glider's nose down in order to maintain the correct towing position. The crude antidote for this problem was to affix a sandbag to the front of the longitudinal under-skid but this had other undesirable 'side-effects' as Bill Denholm and I were to experience. One of many techniques being taught us was that of 'remote release'. On a particularly hot summer's day we were released at some 8,000ft near Didcot to glide back to our Kidlington base. The glider was badly rigged and was well weighted at the nose with a large sandbag. Once in free-flight, it was impossible to trim to an optimum glide angle and our 'pith-ball variometer' told us we were losing height at an excessive rate. I was flying at this

Lysander glider tug (above) and Tiger Moth flight trainer (below).

juncture and contrary to flight instructions decided to fly directly over Oxford hoping for fairly intense thermal activity over the City. I was not disappointed and although not 'soaring' – a virtual impossibility in an Army glider – the variometer showed a marked improvement in our rate of sink enabling us just to clear the near boundary of the airfield and to land successfully, but down wind. Such episodes were by no means rare occurrences and in due course, despite theoretical assumptions. the towing point was moved to the nose of the Hotspur with great success. Many other exciting variants contributed to our flying experiences. Formation flying in free flight using the flaps to maintain position; dive approaches from heights of three to four thousand feet at speeds of 140mph and then low-flying in for a considerable distance, hopefully reaching the airfield before stalling speed was reached; low-level cross country tows; night flying; all with obvious relevance to eventual actual operations.

Imagine then our excitement when an assault on an enemy position was 'mocked-up' in the Oxfordshire countryside. The scenario was as follows. A farmhouse was deemed to be a German radar site from which a glider raid was to secure and remove vital equipment. Five Hotspurs loaded with troops – eight to a glider – towed by Lysanders, landed in an adjacent field, the troops moved in to the attack and simultaneously the Lysanders also landed. The main tow ropes had already been dropped and retrieved by the glider pilots. Each glider had a 6ft tow rope on board and this was attached to the Lysander which then, acting as a tractor, towed the Hotspur across the field to a pre-selected take-off position. Short tow rope then discarded and replaced by a 'retrieved' rope. Troops returned carrying simulated captured radar equipment, enemy routed, exit gliders all of which returned safely to base! Was this a rehearsal for the Bruneval raid subsequently successfully carried-out by Commandos? I shall never know. Suffice it to say that our hopes of early action were rapidly dashed and in the not distant future many of these operational training exercises were abandoned, notably dive-approaching which had been the cause of several deaths, and a single course amalgamating the training schedules of GTS and GOTU resulted. Undoubtedly the unique 'flavour' of Kidlington under the leadership of Group-Captain 'Crazy' Noakes – a pre-war Gloucester Gladiator aerobatic pilot of great repute – was lost when the Hotspur was relegated to a training role only, in the wake of the advent of the Horsa.

There were the usual more light-hearted interludes throughout our stay at Kidlington. A Station Command Order possibly indicates in its terse language some of the nervous tension associated with our day-by-day training. "Due to excessive corrosion on the undercarriages of Hotspur Gliders, promiscuous urination on such undercarriages shall cease forthwith". Mention has been made of our Station Commander, Group-Captain Noakes. In addition to his passion for flying he had a 'touch of the Capability Brown' about him to the extent that the open-sward in front of his personal quarters was to be planted with an assortment of flowering trees and shrubs.

Therein lies a tale. Corporals Pearson and White had been guilty of some minor misdemeanours and were 'sentenced' to plant a flowering cherry tree in front of 'Groupy's' quarters. This was the Pearson who on the eve of D-Day flew in the Coup-de-Main party which captured the River Orne bridge, for which action he was awarded the DFM. But this was far ahead. Tree planting was the immediate task. The two corporals, close friends, were somewhat lugubrious and dour

Yorkshiremen and erstwhile members of the Bulshie-Baiters Club. Left to their own devices they dug a hole almost six feet deep, buried the poor tree up to its lower branches thus reducing it to shrub proportions, and ruining the aesthetic effect desired by the G-C. At the subsequent less than friendly interview, Stan Pearson apologised profusely for his mistake in thinking that the trunk was a tap-root! Did the tree survive or did some unfortunate RAF offender have to dig it out ? I know not but it was certainly the first and last 'fatigue' to be endured by a Glider Pilot on our Course.

Our own Course commander was still Major Willoughby, a fine soldier much respected by the 'other ranks' whose departures from the 'strait and narrow' he controlled with a firm but flexible discipline all too absent in the Army of that time. He recognised us as individuals and treated us as such. One day shortly after we arrived at 102 GOTU, he sent for me, and in a highly informal manner discussed our Ground Training results. At each of the preceding two courses, I had come top and he second. The difference was, apparently, our marks in Morse code, both key and Aldis lamp, in both of which I had been trained in the Signals. Would I coach him – not an order but a request ? Naturally I agreed, Morse code sessions followed and at the end of our course he was top and I was second! Did this influence my subsequent appointment in the Regiment ? I wonder.

Little remains to report from Kidlington. On 6th July, now a Sergeant following the award of my Glider Wings, I had my last ever flight with my great friend Geoff Nelson who was arranging my honeymoon accommodation in Winton, Bournemouth. Hilda and I were duly married on 11th July, the day following her 21st birthday, thus side-stepping any parental intervention and a blissful ten days were spent in Bournemouth where certain local restaurants seemed to have escaped the restrictions of wartime rationing. At this time only thirty or so Glider Pilots had been awarded 'Wings' and these together with our Red berets rendered us so conspicuous that I wore 'civvies' throughout this period. All too soon Service uniform had to be donned as we went our separate ways, Hilda to London as a Lance-corporal clerk and myself to join the first course to fly the Horsa glider but not before a short interlude – mercifully only two weeks – at our Regimental Depot at Tilshead. What a 'dump'. We thought it a bit of a joke when one or the other of us was told to check up if a particular lecture room was still standing following an artillery barrage whistling over our heads, but it was no joke when six months later three corporals were killed by a shell landing short in the Medical Orderly Room! Fortunately much of our time was spent 'off-camp' wandering around England on 'Initiative Exercises' which involved individuals travelling with little in the way of money and no Passes to achieve some hair-brained scheme such as entering Buckingham Palace undetected or penetrating a Top Secret Gunnery Control Room. Geoff Nelson even fell in love on one such exercise and actually married the girl before we departed for North Africa.

Towards the end of July we were finally released from our misery and posted to RAF Brize Norton, then just being set-up as the Heavy Glider Conversion Unit (HGCU) ,where we had our first sight of the Airspeed Horsa. There were eleven of us in this 1st Course, all but one from our original 2nd Course at 16 EFTS all those many months ago, the 'odd man out' being a Lieut Cranmer who, unbeknown to us, had been selected by Lt-Col Rock to carry-out glider test flying at Farnborough, this

Hilda, 1941.

based on an exceptional rating throughout his flying training. In fact it is highly probable that he was the first ever pilot in England to fly a delta wing aircraft, a wooden mock-up towed as a glider in order to evaluate its handling characteristics. He stayed at Farnborough for the duration of the war, returned to the Glider Pilot Regiment as a Captain with flying log books the envy of us all – Spitfires, Typhoons, Tornados and a host of other 'exotic' aircraft filled the pages as he progressed from Glider Pilot to a highly competent all round Test Pilot. He even included rocket assisted glider take-offs in his wide experience. However, although he was the only member of this Course to graduate to such dizzy heights, other individuals are worthy of mention. His fellow officer, Lieut Robson was seriously injured in a flying accident and was 'grounded'. Of the remainder, all sergeants, Doig flew on the abortive Norwegian 'Heavy Water' raid on 27th November '42 and was killed, probably by the Gestapo; Nelson, White and Denholm were all killed in action; Pearson and Guthrie flew together on the Coup-de-Main operation the night preceding D-Day and participated in the vital capture of the River Orne bridge for which Pearson was awarded the DFM and Guthrie the Croix de Guerre and I was fortunate to be appointed to an HQ post which enabled me to be on the 'inside' for the planning of the Sicilian operation and to a lesser extent, Arnhem. Regrettably I have no knowledge of the remaining members – Taylor and Torrance – who disappeared from my 'view'.

But what of the centrepiece of this section of my story, the Horsa glider. My immediate reaction on first sight was one of disbelief. Could such a 'monster' once airborne fly without engines ? It sat well above the runway on its tricycle undercarriage necessitating a five rung ladder to get inside!

Head-on the 88 foot wing-span was impressive whilst from the side the 67 foot long fuselage terminating in a high tailplane and large rudder was more than a little daunting. I need not have worried. Airspeed had designed and developed a magnificent aircraft, well equipped with massive flaps resulting in excellent landing characteristics. These together with the wheel brakes operated by compressed air gave further control on the ground, all in all belying the misplaced description of it as 'a flying coffin'. No such thing – a magnificent aircraft. It weighed about 3½ tons light and 7 tons fully loaded with either 25 – 30 troops, a jeep and a 6 pounder anti-tank gun or other similar loads. Its one serious shortcoming was off-loading jeeps, guns etc. After landing, the tail which weighed nearly half a ton, had to be removed, no mean feat as the so-called quick release bolts were anything but 'quick', metal troughs were then located at the rear and the load then driven out! Not good when under enemy pressure. The subsequent Mark 2 version overcame this problem by having a hinged nose but unfortunately it was 1945 before this version was available and then only in limited numbers in time for the Rhine crossing. This then was the glider which was to be the principal 'work horse' for the British Airborne Divisions although it was in due course exceeded in size by the Hamilcar, surely a real monster, with the capability of carrying a 7 ton tank or self-propelled gun and an all-up weight of 15 tons. The 'Ham' as it was known never figured in my flying experiences.

Understandably, Bomber Command were reluctant to release operational aircraft for mere towing duties and as a consequence Brize Norton was equipped with the Whitley, clearly underpowered for the task of Horsa towing particularly on full load. It was eventually replaced by the American Douglas

Dakota and the Albermarle, an under-performing aircraft not wanted by Bomber Command. There can be no doubt that the Whitley aircrews had a more daunting task than those at 'the other end of the rope'. Meanwhile personal disaster struck. After one trip on 28th July, I was grounded with severe sinus trouble and, had it not been for the shortage of trained Glider Pilots that would have been the abrupt end of my flying career. I had lost all control of my balance to such an extent that I had to be tied into bed to prevent me rolling out onto the floor! Many days elapsed in the Sick Quarters at Brize and despite powerful inhalations and searing drops in the nose, no improvement was forthcoming. An ambulance whipped me off to RAF Wroughton Hospital where an RAF Squadron-Leader Surgeon, with the inevitable nickname 'The Butcher', pierced through my nostrils into the sinus cavity – excruciatingly painful. I returned to Brize, a few days leave and resumed flying training on 17th September feeling somewhat lonely as my original course had returned to Tilshead. But one episode is worth placing on record. I had 'gone solo' on the Horsa and flying as 1st Pilot shortly after, on my approach glide at about 500 feet the glider suddenly flicked into a half-roll and plunged earthwards. Fortunately I had sufficient height to correct the situation by aileron and rudder and we landed safely but more than a little shattered by the experience which was certainly not in the Training Manual. The tractor driver, as always on hand to tow us back to the end of the runway, gesticulated madly and retrieved us back to the hangar where, on inspection we saw the cause of our troubles. I had approached on half-flap – standard practice – the flap on the starboard had collapsed due to a fracture in the metal lever mechanism from the air cylinder to the flap. An Official Enquiry ensued at which I was cleared

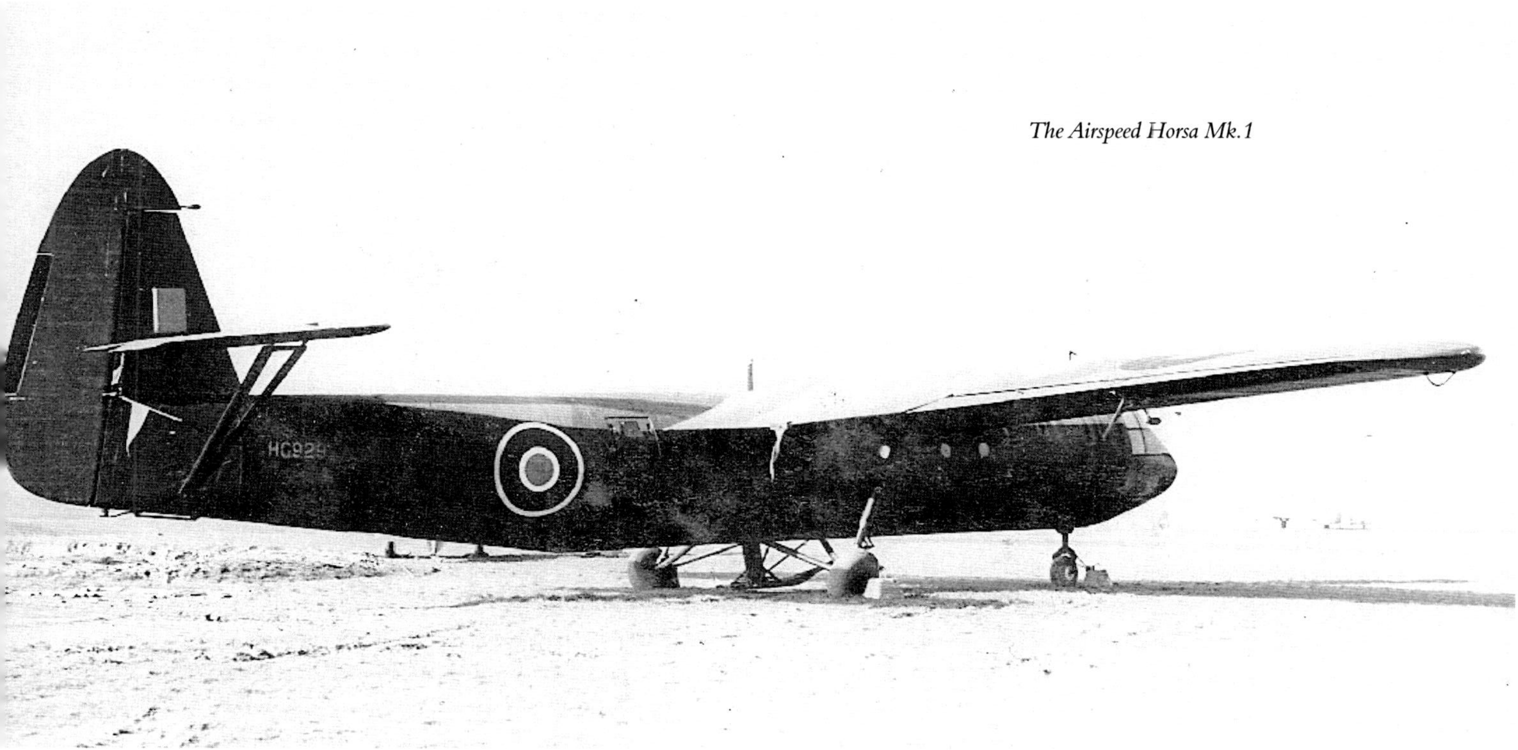

The Airspeed Horsa Mk.1

of any malpractice and was congratulated on my safe landing. There was a sequel to this episode on 2nd November '42 when, at Netheravon a demonstration was laid-on for a dignitary from the War Office with, I seem to recall, the unlikely title of 'Director of Air, War Office.' (?) I was again flying as 1st Pilot but with a full load of some 25 troops. At about 500 feet there was a ' carbon copy' of the flap episode, half-roll, but this time in the other direction. Port flap had failed. Sufficient height to knock-off half-flap, level the glider and make a flapless landing. All was well though troops somewhat shaken. Ordered to appear in front of visiting dignitary who congratulated me on an excellent demonstration of *Hotspur* (sic) flying. If he didn't know the difference between a Hotspur and a Horsa, I had grave doubts as to our future! I thanked my lucky stars however for my previous experience in training albeit by pure chance. I have digressed but have no reason to return to Brize Norton as my training had concluded, successfully, on 27th September, coinciding with the loss of our Commanding Officer, Lt-Col Rock in a night flying accident at Shrewton, a loss from which the Regiment was never to recover. Thanks to my sick interlude I was now posted direct to Southborne, Bournemouth where I rejoined my old friends who were established in Crawford's Cafe, now to be our billet and Sergeants' Mess for some weeks to come whilst we set up a Glider Pilot Battle School. Lucky me! I had missed-out on their intervening three weeks enforced residence at the dreaded Tilshead!

It can hardly be said that my Battle School experiences contributed in any significant way to my subsequent Army career but, unbeknown to me, it was the last truly carefree period in my Service life and was terminated by an appointment which

Glider tugs: the Armstrong Whitley (above) and the Albermarle (below)

undoubtedly represented a turning-point in my personal development in more ways than one.

I had not only rejoined my previous mates but had caught up with many of the 1st Course which had preceded us at Burnaston. What a wonderful collection of individualists the like of which I never again met in any walk of life. The initial task in the development of a Battle School was the construction of an Assault Course as 'popularised' by 'Monty', So off to nearby Lymington – 46 Division if I recall correctly (?) – to experience the delights of an existing Assault Course – an horrendous experience as far as I was concerned. As a non-swimmer crawling through a flooded culvert with my nose barely above water and struggling to keep my rifle dry, was no fun but at least it inspired me to construct a similar hazard at our own Battle School near Hurn. It was dubbed 'Waldron's Widdle '. We were nominally at least under the control of Lieut Iremonger, himself an individualist of the highest order, destined to fall foul of the local Garrison Authority. It happened thus. He decided to test us with a night attack on the nearby Pokesdown Railway Station. He was to 'command' the defending forces and we were left to 'elect' our own 'officers' for the assault force. We were also given a free hand in drawing blank ammunition and other relevant offensive items from the magazine and this we took full advantage of to the extent that one of our number 'nicked' two canisters of a noxious yellow smoke. As part of the attacking force I assembled on the beach at dead of night. Faces were blackened, Commando style woolly balaclavas in place of berets or helmets and up the crumbling cliffs we swarmed. First objective, top of the cliff, attained without 'loss of life'. Our advance took us through a bungalow area populated by retired citizens who were convinced that this

Operation 'Coup de Main'
6th June 1944:
Horsa gliders successfully find their landing zones in the dark to secure a strategic bridge.

was a German invasion. Undeterred we moved on to the 'T' junction opposite the Station at which point the 'Iremonger force' took-up defensive positions in the Station Booking Office and refused to budge thus forcing us into an all-out assault. One section was dropped over a parapet wall onto the railway track with instruction to signal by Very light when in position. The afore-mentioned smoke canisters were then 'released' at the 'T' junction to provide cover for a frontal attack and as the 'track' party attacked along the railway line we went in with fixed bayonets – strictly against all military rules – and to the consternation of waiting passengers, the objective was achieved. All adjourned to the adjacent pub to down a pint or two and savour our success. Outside, on a dank October night, the trolley bus service had been brought to an unscheduled halt by the 'pea-souper fog' emanating from our canisters. The local Garrison Commander was not amused particularly as he had not been informed of the intended exercise and in all probability many local citizens did not sleep easy in their beds that night.

At about the same time, on a glorious Sunday morning, we, together with a Company of the Black Watch, were asked to test the defensive positions of the local Home Guard around Hurn aerodrome. All went well until we came up against a substantial brick barn, one of the Home Guards' forward defensive positions. We launched an attack which the 'Umpire' refused to recognise as effective. Staff-Sergeant Taylorson was our 'elected officer commanding' for that day and I had been assigned to the 2-inch mortar, a notoriously inaccurate weapon. I was ordered by Tommy Taylorson to lay-down smoke to cover our grenade thrower, Sergeant 'Dinger' Bell, to reach the barn door, open it and 'pop' a grenade inside. By misfortune we had no practice grenades so a plastic 69 was used – the real thing –

Operation 'Coup de Main'
6th June 1944.
Another successful landing, after a six-mile glide in the dark, secures strategic bridge.

exit Home Guard slightly bemused but generally unharmed, objective achieved and on to the Home Guards' last stronghold, an old windmill. Smoke was once again required for covering the final assault. First smoke shell somewhat short – up twenty shouts Tommy – an impossible accuracy for this weapon. I placed a salvo clean through the roof of the mill, fire broke out in the tinder dry grain therein stored, Home Guard very rapid exit and mill substantially damaged if not burned to the ground. Oddly enough in 1946 I met a civilian in an RAF Officers' Mess at Brize Norton, who as the owner of the Mill regaled me with this story whilst I maintained a discreet silence. Such continued silence is no longer necessary as I am surely now immune from prosecution for wanton arson!

A telex from HQ, now at Old Kiwi Barracks, Bulford, brought an abrupt end to this phase of my Glider Pilot service. Major Willoughby, my old Course Commander and now acting CO of 1st Battalion (our new CO, Lt Col Chatterton was temporarily absent on glider training), required my services at Headquarters. It was an order which had to be obeyed. Reluctantly I left Southborne and was never again to enjoy the camaraderie of the Denholms and Nelsons, the Pearsons and Whites with whom I had embarked on my unforeseen training as a glider pilot in January 1942. Unbelievably, thinking back over my many varied and challenging experiences although it was November it was still 1942.

CHAPTER 8

Intelligence by Decree

The telex from Major Willoughby – acting Commanding Officer in Lt Colonel Chatterton's absence on his glider flying course – appointing me Staff Sergeant (Intelligence), was not an invitation but an Order which brooked no refusal. Having already been promoted Staff Sergeant, neither an increase in pay nor an improved rank provided any incentive to leave my 'Squadron mates' so it was with ill-grace that I packed my kit and awaited the truck to transport me to 1st Battalion H.Q. at Bulford near Amesbury. At my farewell evening with 'the lads', I parodied a passage from Shakespeare's Twelfth Night – Some are born intelligent, some achieve intelligence and some have Intelligence thrust upon them. It was certainly in the latter category that I placed myself and mused that I should never have coached the Major in Morse Code. Many years later he averred that this had nothing to do with my appointment. Little did I realise at the time that this was yet another 'milestone' in my personal development and one which would have a significant and lasting impact not only on my Army life but also on the direction which my distant post-war career was to take.

My first impression of HQ did little to raise my spirits. Old Kiwi Barracks, Bulford, was a First World War set of buildings long overdue for demolition but 1st Battalion's temporary home until such time as a purpose-built HQ and Depot was completed at Fargo near Larkhill and in close proximity to

Stonehenge. We had been promised a 'luxury camp' by Chatterton in one of his typically flamboyant and shallow utterances. To describe a featureless Nissen hutted camp in the middle of nowhere as 'luxury' was some overstatement! From the southern end of the camp, Stonehenge could be seen on a damp and misty winter's morning monumental and brooding in the near distance, an impression etched deep into my memory and destroying any interest in its unquestionably fascinating past. These two locations, Bulford and Fargo, were to be my 'home' for the six months prior to our departure to North Africa in April 1943, a posting hailed with great relief and even enthusiasm by all concerned. Surely nothing could be worse than a drab life on Salisbury Plain in the middle of winter! I should be thankful that throughout this period I was attached for flying to the Experimental Unit at Netheravon. At a time when flying hours were almost non-existent I was actually airborne on fourteen occasions, by no means sufficient to maintain one's skills but better than nothing.

In sharp contrast to the drabness of our living quarters were three outstanding and wonderful characters, two of whom were to become life-long friends – Regimental Sergeant Major Michael Briody and Staff Sergeant (later Warrant Officer 2) Joe Lowe i/c Battalion Orderly Room. The third member of the trio, WO2 Ron Spathaky attached to us from the Army Education Corps, had a profound and lasting effect on my political thinking even though I lost touch with him shortly after my demob in October 1946. As their presence will run like a thread through the remaining chapters of this book, a 'thumb nail' sketch of each of them seems appropriate at this stage.

Michael Briody, the 1st Battalion's RSM, had joined the Glider Pilot Regiment from the Irish Guards where, as a Regular soldier, he had served as a Warrant Officer 2. Surely no finer man ever joined our Regiment. In order to sum-up his many virtues I can do no better than quote from the obituary notice which I wrote for our Association magazine – the Eagle – when he died on 6th April 1991. "It is surely a rare occurrence for an RSM to engender such genuine and deep affection and can only stem from his unique qualities. He enforced discipline without humiliation; he maintained his personal authority but was never aloof; he radiated humour, sincerity and friendship but never courted popularity as an end in itself; he was possessed of a great inner strength which undoubtedly stemmed from his unshakeable faith. He was indeed a 'man for all seasons', respected by officers and other ranks alike...". Such was the nature of the man, senior to me in rank, with whom I served for three years but whose friendship I was privileged to enjoy until his untimely death in 1991.

Joe Lowe was a very different character. He had transferred from the Lancashire Fusiliers to train as a Glider Pilot but in order to 'survive' his RAF medical examination had called upon his memory – almost photographic – to conceal the fact that he was colour blind. Alas, his Nemesis overtook him when flying in a Tiger Moth at Clyffe Pypard he failed to respond to a red Very light, thinking it was green! Abrupt end of flying career. However, his 'loss' was the Regiment's gain as had been duly appointed as Orderly Room Sergeant (Chief clerk in civilian terms) by Lt Col Rock some months before I arrived at HQ, by which time Rock had been killed and Chatterton had taken-over. We served together throughout the rest of the War, became in due course the longest serving inhabitants of HQ and some

57 years later are still close friends. It speaks for itself. Joe was a highly efficient organiser of the Orderly Room but above all totally discreet in all matters where confidentiality was required. My own duties were by their very nature shrouded to varying degrees in secrecy but I could at all times share many of my problems with Joe, take him into my confidence when I felt in need of a 'second opinion' and thus test my own judgements – essential when feeling isolated by a series of ineffectual Intelligence Officers who, in the main, left me to my own devices.

The last of this trio of friends and colleagues, Ron Spathaky, was cast in an entirely different mould. An ex-schoolmaster peering out from behind his steel rimmed spectacles, apparently bemused by his position as a Warrant Officer in His Majesty's Armed Forces, he was in many ways the epitome of the absent-minded professor. A dedicated Marxist and a member of the Communist Party, he enthused me with visions of a Classless Society as portrayed in the writings of Marx and Engels, introduced me to an analytical reading of history and persuaded me to join the Communist Party, a short-lived experience terminated early in 1944 when I was disillusioned over the 'Second Front Now' campaign then being propagated by 'The Party'. On reflection I have wondered whether the appeal of the highly disciplined CP wasn't some kind of replacement to fill the void created by my leaving the Exclusives, an equally disciplined and doctrinaire organisation ? Despite my brief membership of 'The Party', Ron's clandestine educational activities in the field of Marxism and Dialectical Materialism have influenced me throughout all my post-war political activities within the Labour Party. Even now as I read Eric Hobsbawm's brilliant four volume history covering the period

1789 to 1991, I revel in the insights that his Marxist thinking brings to bear on the ebb and flow of human affairs. But I have digressed. I had but one brief post-war encounter with Ron Spathaky when he sought to persuade me to rejoin the Communist Party, but I firmly rejected his overtures and we were never again to meet. Sadly I realised that our 'friendship' was wholly dependent on political affiliation!

The way in which each of this 'trio' was to impinge on my new life as Intelligence NCO was unknown to me as I buckled down to the initial task of creating some administrative order out of the chaos which I found in the shape of two 'secret boxes' stuffed with sundry correspondence stretching back over the preceding year or so. My Civil Service experience came to my rescue. I painstakingly sorted all the many letters into date order, posted them into a 'Communications Received Register', assessed their 'Subject' content and then created the necessary files suitably coded. The fly-sheets within each file contained a summary of its contents and two filing cabinets were soon crammed full with the product of this past correspondence. The letter securing 'pride of place' both by date and content was of historical importance. Over the Prime Minister's signature it issued instructions to his Chief of Staff to set-up a gliderborne force. I coined the near blasphemous expression – 'In the beginning was the Word and the Word was Churchill'. Meanwhile I had been instructed in the mystique of 'Security Classifications' which involved segregating the files into one of three categories – Confidential, Secret or Top Secret. To over or under classify letters and files was apparently a 'crime against humanity', resulting in highly subjective judgements having to be made, but as I was 'Judge and Jury' on such issues it never caused me any loss of sleep. Once this initial task was completed

I attended rudimentary training sessions, probably at Netheravon, on map-reading, identification of aerial photographs and the 'art' of writing Army Operation Orders, an excellent discipline which served me well in Report Writing in civilian life. Inadequate as this training was, I had reasons to be extremely grateful for such basic knowledge when I was 'chucked in at the deep end' in the planning phase of the Sicilian operation, a story yet to be told. Such were the modest beginnings of an 'Intelligence' career which was to last until the winding-down of the Regiment in 1945 by which time I was Operations Officer with the rank of Captain. During the whole of this period, though in theory serving under the direction of an Intelligence Officer, in practice I was left almost entirely to my own devices and regrettably, can recall but one officer who made any significant contribution to the work of the Intelligence Section.

The tedium of HQ life at this time was not relieved by our move to Fargo. Ron Spathaky was in his element as ABCA (Army Bureau of Current Affairs) lectures were focussing on the Soviet Union our ally since June 1941 when Germany had invaded the country and a small group of us continued to have 'extra-mural' instruction from him into the esoterical world of Karl Marx. Far from having any adverse effect, it inculcated into us a more positive attitude to our involvement in World War 2. But for this would I have reacted as I did on receipt of a Top Secret letter in the Intelligence Section, signed by no less a dignitary than Bomber Command Harris, in which he proposed to limit post-training Glider flying hours to a derisory 7 hours per year! Needless to say had such a policy been adopted it would have been the death of the Glider Pilot Regiment in more ways than one! I broke confidentiality by discussing this

Top Secret document with Ron Spathaky and, not without fear and trepidation, I agreed to arrangements for me to meet D.N.Pritt K.C, an ultra-left M.P. The meeting duly took place in his Middle Temple Chambers. He was a formidable figure and a brilliant K.C. possessed of impressive forensic abilities which enabled him to study the brief and assess its significance in but a few minutes. His immediate reaction was that a matter of such importance warranted the placing of a Parliamentary Question to the Prime Minister but recent research has revealed that he in fact wrote to Churchill (I have a copy on file) who in turn called for a report from his Secretaries of State for War and Air. At the conclusion of the interview he thanked me for my intervention, commented on the risks I had run and then, fearing that his Chambers were under surveillance, instructed his Clerk to lead me through a labyrinth of back alleys to the anonymity of Fleet Street and the Strand. A chance meeting with a senior Air Force Officer many years later revealed that Churchill had in fact called Harris to a meeting at which, following much discussion, he was given an ultimatum – tow gliders or resign. The crisis was resolved, the necessary flying resources were allocated to the Regiment and in due course two Wings were trained, one of which served with the 1st Airborne Division and the other with the 6th. Whilst all of these to-ings and fro-ings were taking place behind the scenes – unknown to me at the time – I had departed with the advance party to North Africa to help plan and participate in the ill-fated Sicilian fiasco.

It was a grey and drab morning on 13th April 1943 when a tender ferried us out to a troopship – the Niew Holland – lying off-shore at Gourock. Following instructions from 'on high' we had divested ourselves of our much cherished 'Wings', our

The Waco glider (later renamed the Hadrian) ...
...and the view from inside its cockpit.

badges of rank, hidden our red berets at the bottom of our kit-bags, donned badgeless forage caps and were duly treated as infantry privates with all the irritations and privations that that implied on an overcrowded troopship. This was all in the interests of Security! After a fractious voyage we disembarked at Oran and were then transported through the Algerian countryside, further and further South – we actually saw a signpost to Timbuctoo! – and then suffered the further humiliation of being incarcerated in a newly constructed prisoner-of-war camp, manned by American guards. This, followed by the most torrential storm that washed tentage and personal kit away in its wake, came close to creating a mutiny, but wiser counsels prevailed, not least due to the camaraderie of the US troops, games of baseball, belated sunshine and the exotic joys provided by a typical French bar in nearby Tizi. At this time HQ had departed to the French colonial town of Mascara, leaving me to rejoin my old Squadron, a prospect that did not displease me. But in the upper echelons of Command 'things' were moving and as a consequence a small detachment, happily including me, was moved to the vicinity of Froha, an Arab village to the south of Mascara. Rumour had it that an airstrip was being constructed for us nearby, by the Americans and, mercifully for once, 'rumour' was accurate. We had once again donned our berets, Wings and badges of rank and were unfazed when we found ourselves digging regulation depth latrines, erecting a large marquee to serve as our Sergeants' Mess and sundry other construction duties. Flying was once again in the offing though we were unaware that the gliders we were to fly, American Wacos (eventually re-named Hadrians), were still crated-up at an airfield near the coast and shortage of riggers was hampering their assembly. In due course Chatterton (to be known as 'George') arranged for Glider Pilots to be detached and trained to assemble and rig these gliders and thus speed-up their availability for training and eventual operations. Once again I was to be uplifted from the 'bosom of my friends'. A jeep arrived from Mascara and I was ordered to report forthwith to the CO, Lt Col Chatterton. One redeeming factor was my reunion with my close friend Joe Lowe, now well established in a Mascara hotel stripped of all but the very barest of essentials, where 'George' had set-up his HQ. Typical of Joe and his organisational flair, the Orderly Room was functioning as if it had been established in North Africa for many a year.

The ensuing two to three months were certainly the most bizarre of my Army life – perhaps unbelievable would be a more appropriate description. A newly appointed Intelligence Officer – Peter Scott-Malden- a man of considerable ability had, whilst still in England, been posted to Matlock, Derbyshire, on a three months Intelligence Course and there he had remained as we embarked for North Africa. George Chatterton had belatedly realised that he had no Intelligence Section and therefore plucked me out of my Squadron activities. He hardly knew me as an individual and I had certainly had little contact with him after his return from Glider flying training as I continued to report, as and when necessary, to his second-in-command, Major Willoughby, who had appointed me in the first place. Thus, at a critical time, facing as we were the run-up to the first ever large-scale gliderborne operation – the tragic Norwegian operation had involved only two gliders – we were without an Intelligence Officer and, ,unfortunately George Chatterton proved to be woefully short in this field. First-rate pilot he certainly was as I can testify from personal experience, but in affairs 'military' he showed his lack of Staff College training time and time again and had little to offer either in tactical or strategic thinking. Again I speak from personal observation. In the event he suggested that his 2nd Pilot, Lieut Peter Harding who had a purely nominal role as Assistant Adjutant (no such post figured on our Establishment) should serve as Intelligence Officer at this time, a suggestion which I forcefully rejected. To my amazement he acceded to my request to 'act on my own' and as a consequence I was duly appointed to the XO Planning Committee, the highest echelon in the Operational Planning set-up, chaired by Major General Hopkinson and, I say with tongue in cheek, with a mere Major as the 'tea-boy'! What an experience! It was to be a joint operation. The majority of the towing aircraft were to be supplied by the American 51st Troop Carrier Command-under the command of Colonel R.A.Dunn- a factor that had a overriding influence on route planning and an unforeseen disastrous effect on the outcome of the whole operation. Other members of the XO Planning Committee were the Brigade Commanders and the senior Divisional Staff Officers – GSO 1's – of whom there were two, one known as (Q) and the other as (A) with special responsibilities for, broadly speaking, 'supplies' and 'deployment' respectively. These officers held the rank of Lieut Colonel.

Two pitfalls must now be avoided as I write my 'worm's eye view' of the ensuing events. I must not be tempted into writing an analysis of the Sicilian operation which was aimed at securing a spring board for the invasion of Italy. That has been done many times by far more competent military researchers and historians than I can ever aspire to be. Nor having read such accounts must I allow them to colour my own fragmentary and possibly inaccurate memory. Indeed, although most of the details of the XO meetings are lost in the mists of time, particular memories stand out with remarkable clarity, probably due to the fact that throughout the ensuing years, Joe Lowe and I have reminisced over those distant affairs time and time again. His companionship was vital to me at that time allied to the fact that in him and him alone I could confide Top Secret information knowing that such confidentiality would never under any circumstances be betrayed.

Contrary to Chatterton's much publicised account of his clash with General Hopkinson -" I was ordered to fly or resign" – I remain firmly of the opinion that whilst we would all have preferred a day rather than a night landing, our CO would never have even contemplated the exclusion of gliders from the operation. He was too conscious of his own advancement to consider such a possibility. A night landing was, understandably, obligatory from the American viewpoint due to the vulnerability of the Dakota to flak – no protecting armour or self-sealing tanks. My rudimentary knowledge of the interpretation of aerial photographs proved valuable as Chatterton had no experience in this 'art form'. Squadron Leader Lawrence Wright, an RAF boffin highly experienced in glider 'technology', supplied load, flying and glide data for the relatively unknown Waco glider which we were to use. The final plan both as regards the three Landing Zones (LZ's) and the approach over the sea towards the Cape Murro di Porco, was a compromise – what Battle Plan is not? – providing the best options for gliders to land in proximity to the Military Objectives – the securing of two vital bridges – whilst maximising the safety of the Dakotas vis-a-vis known flak concentrations on the Cape Murro di Porco. Two of the selected LZ's were at differing distances from the coast and it was therefore decided to fly in two streams, one at 1400ft for the nearest LZ and the other at 1800 ft for the more distant one. Meteorological information, such as it was, indicated an off-shore wind of some 5mph and this was taken into account when the glide paths were calculated. As a mere Staff Sergeant my opinions were neither sought nor permitted – like Victorian children- S/Sgts should be seen but not heard – as it was clearly an Army dictum that intelligence was directly correlated with rank! Nevertheless I had very clear and firm convictions that not only would the Operation be a total disaster but also I would be unlikely to survive the debacle. Although unable to give vent to this gloomy outlook I felt a responsibility to place on record certain facts and opinions for future reference. My ever reliable friend Joe Lowe now comes back into the story and but for his willing assistance there would be little left to recount.

Two key dates were fast approaching – 9th July the date set for the Operation and the 11th my first wedding anniversary. As I have already stated this was the first Allied mass airborne operation using airborne troops and gliders. Everybody concerned in the planning was thus working as it were from a blank sheet of paper, there being no previous experiences on which to call for guidance. I therefore envisaged my successor being placed in a similar situation in the absence of any planning records written from the glider point of view. It was also obvious that in hindsight recriminations would abound, the quality of decision taking would be challenged and in retrospect we would all look like congenital idiots. To eliminate at least some of this probable backlash I decided, with Joe's help, to compile a record of problems faced, decisions taken and the rationale behind such decisions as seen at that time. After a lapse of 56 years I am unable to recall this document in detail but suffice it to say that it in all probability dealt with the characteristics of the WACO, day and night flying techniques, ground organisation, mass take-offs, rendezvousing in the air, flight formations, identification of LZ's, release heights etc. The document, painstakingly typed by Joe Lowe in his role as Orderly Room S/Sgt, was presented by me to Chatterton shortly before the Operation and was received gratefully with the comment, "We mustn't let the Divisional 'Gee ones' (GSO 1s) see this Staff," and it was then consigned to the Battalion Safe, the haven for all Secret documents. A few days later I left a letter to my wife for Joe to post in the event of my non-return but thankfully this proved to be unnecessary.

Throughout all the preceding activities I had managed to clock-up some flying hours in the Waco, both day and night, the latter in the company of George Chatterton who, whatever his other shortcomings, was a first-rate pilot. I also flew the last glider combination in a mass take-off under the code name 'Eve' in order to observe the effect of dust and sand on visibility. I reached a height of 400ft before seeing my tug, a Dakota. Such was the problem which all glider pilots faced day by day as they logged precious but all too few flying hours in the Waco, a glider with widely different flying characteristics from the Horsa on which we had all trained. Towards the end of June, 1st Airlanding Brigade was uplifted by gliders over a distance of 650 miles to Tunisia where six airstrips would act as the launching pads for the forthcoming operation. And so to Exercise 'Bigot Husky', the code name for the invasion of Sicily, as recorded in my flying log on 9th July 1943.. My forebodings were all but fulfilled at the outset. Having assisted at the briefings I hastened to my glider at Airfield 'F' where I was to be towed by an Albermarle of 296 Squadron. My co-pilot was Sgt George Harris, a real stalwart to whom we probably owed our survival. Our Waco had been declared u/s with defective ailerons but having been involved in the planning and subsequent briefings, it was unthinkable for me to remain on the ground. I remonstrated with the American Staff Sergeant i/c ground crew who wept as I insisted on taking-off, in retrospect a foolish and irresponsible decision which put my live-load (Border Regt) at risk. The only way in which to maintain a reasonably level attitude was to pick-up a dropping wing by yawing the glider using excessive rudder, a tiring and uncomfortable manner in which to fly a glider! It also made it difficult to keep station on the tug as darkness fell and the pinpoint light on the tail of the Albermarle was the only indication of its position. The mesmerising effect of watching this elusive light as it moved left and right, up and down, and in and out of my vision finally caught up with me somewhere over Linosa.

I temporarily 'lost it' and but for George Harris's presence of mind and expertise we would probably have 'ditched' somewhere in the Mediterranean. Control was regained and at the planned 1400 feet we made our approach run towards Cape Murro di Porco unaware that an off-shore wind of 25 – 30 mph was blowing, flecking the waves with 'white horses'. At this crucial stage of the flight, communication with the tug had perforce to be cut-off as it was dependent on a field telephone linked to the Albermarle by cable. Failure to detach the phone from the cable before casting-off could have had disastrous consequences as the still connected instrument would have leapt forward possibly smashing the perspex of the cockpit cover. As it was immediately apparent that we were going to be forced to 'ditch', our live load was instructed to prepare for this emergency. As the sea 'came up to meet us' I managed to touch-down tail first. All our personnel escaped unharmed and as the cockpit filled with water George and I managed to force our way through the perspex and thus onto a wing. I alone was a non-swimmer so remained clinging to the wreckage from which I was rescued some seven hours later. All the others swam ashore. I collected a minor wound in the groin, in all probability from machine gun fire ricocheting off a wave but looking much worse after long immersion in the sea. I was never to see George again save for a chance meeting in Manchester some years after the end of the War. Sadly, he had to be invalided out of the Glider Pilot Regiment, the reasons for which were never revealed. An official debriefing report states that my glider released 'at about 6½ miles off shore and landed 3 – 4 miles off shore'. Whence came such information I know not – certainly not from me – as it is utter nonsense. It makes me wonder how much more spurious information is contained in such 'official'

Dakota glider tug.

records, bedevilling the work of would-be researchers? What is clear, beyond all doubt, was the scale of the disaster. Of the 143 gliders involved, only 54 reached land and they were scattered over a distance of some 25 miles.

But to return to the activities of the 10th July -my wife's birthday and the eve of our wedding. The small craft that rescued me took me to a Polish vessel, the Sovieski, and thence to Malta where I was placed in a hospital staffed by Nuns and from where I sent a cable to England telling my wife that I was 'still alive and kicking'. It still amazes me that such communications were possible from such an embattled Island. I was but a few days recuperating from my experiences when I was whisked away to the airport where Lt General 'Boy' Browning's personal Dakota – by no means basic in its accommodation – was waiting to fly me back to North Africa. On landing, I was met by the afore-mentioned GSO 1's and instructed that at a forthcoming Enquiry into the 'debacle' I was not to heap blame onto the Americans. Apparently an Anglo/American full scale row was brewing! I made it quite clear that if called to such an Enquiry I would testify according to my own judgement of the events. I was in fact never called but returned to the Regiment where I was briefly 'sent to Coventry' for my supposed contribution to the disaster. A promised BEM and Immediate Commission never materialised, though a prized possession is a letter of thanks from 'Monty'. Strange as it may seem, none of this unduly fazed me. I had absolutely no ambitions to be commissioned, much preferring to stay as an NCO with the wonderful group of 'lads' who constituted 1st Battalion HQ. Likewise a medal award was of little consequence in the aftermath of the loss of so many pilots and, in particular, Geoff Nelson, my long-standing friend from those far-off days at

16EFTS, Burnaston. He had been killed flying into a cliff, one of 57 glider pilots to lose their lives in the airborne invasion of Sicily.

Meanwhile the remainder of the 1st Battalion arrived in North Africa under the command of Major Willoughby who, following a confrontation with Chatterton was ordered out of the Regiment – a sad loss of an excellent officer. Needless to say the reasons behind this action were unknown to us lesser ranks but rumour had it that it concerned the inept handling of the Sicilian operation. Information which has come my way in recent years indicates that the contentious issue of release heights may well have provided the 'flash point' in exchanges between the CO and his erstwhile 2 i/c and specifically why they were not raised when the American Commander informed Chatterton that an off-shore wind of 35mph was blowing off Sicily and not the gentle breeze of 5mph as included in our release height calculations. Shall we ever know the truth surrounding this key issue ?

The final twist to an increasingly bizarre story concludes the Sicilian affair. The future of Airborne forces was clearly in the melting-pot at this time, and in particular Chatterton saw a direct threat to the survival of the Glider Pilot Regiment and thus by implication, his own future. What should be done ? The case for survival needed to be presented at the highest level in the UK. and Lt Harding obliged by 'stepping into the breach'. He convinced Chatterton (by this time he was his Adjutant), himself no stranger to plausible stories, that for reasons best not stated here, that he could well 'get the ear of the Old Man (Churchill)' and he was forthwith flown home in an Albermarle with, as it later transpired, the report which Joe and I had produced on the eve of the Operation. Suffice it to say that by the time I returned to the UK some six months later, Harding was a Major, G2 (Intelligence), that the Waldron/Lowe report had been reissued over Harding's name and that, as far as the powers that be were concerned, he had been the Intelligence Officer (Gliders) for the planning of the Sicilian operation! Undoubtedly this was done with the full backing of Chatterton. Such are the fortunes of war! Interestingly, despite the efforts of researchers at the PRO Kew, this document has never been unearthed. If found it could well resolve the release height question referred to in the preceding paragraph. In the absence of such information, I shall continue to assert that the release heights of 1400 and 1800 feet respectively were as originally planned, were never raised in response to the revised meteorological information, that this was a key factor in the near disaster and a high level cover-up has deliberately obscured this matter.

I make no apologies for the foregoing digression into the 'present' but now return to our HQ at M'saken, near Sousse and the aftermath of our operational experiences. Of prime importance at a personal level was my learning to swim thanks to the efforts of our recently arrived PTI Sergeant J.L.S.Booth, an old friend from Fargo days. He had been attached to us from the Army Physical Training Corps – a first class athlete, an entertaining companion and yet another life-long friend. I soon developed into a reasonably long-distance swimmer and it is thanks to John that I subsequently avoided drowning in the Rhine during the retreat from Arnhem. Yet another key piece in the jigsaw of my life. A rest camp at Hammamet did much to restore our spirits whilst we were waiting, unsuspectingly, for our next assignment which to our dismay, turned out to be a seaborne operation aimed at capturing the port of Taranto in the south of Italy.

By now it was early September and Chatterton having returned to the UK in August, John Place had been promoted to Lieut Colonel i/c 1st Battalion, and he in turn had promoted John Blatch to Major as his second-in-command. Secrecy was such that apart from a handful of senior officers no one knew what was in the offing even when we were trucked to Bizerta and found ourselves boarding the Princess Beatrix, a specialist Combined Ops vessel. Once afloat we were informed that Taranto was the objective and that we – the glider pilots – would function as an infantry battalion under the command of 1st Airborne Division, news that was not well received 'below decks' where doubts were cast on the mental state of 'Higher Command'. What sense could there possibly be in exposing trained pilots to yet further losses in return for a minimal contribution as under-trained infantry? As we approached Taranto we were transferred to the ship's landing barges, thus heightening the impression that were facing a do or die beach landing possibly, against stiff resistance.

Nothing could have been further from the truth. We cruised into the harbour, tied-up alongside an elevated quay, climbed the stone steps to the top where the Mayor and Corporation of the City awaited us, full of apologies that one of our ships, *HMS Abdiel*, had hit a mine (German of course) with the loss of 130 officers and men of 6th Para Battalion. No remnants of the German forces remained in Taranto so we enjoyed a kind of victory march through the city to the countryside beyond where once again, as at M'saken, we encamped in yet another olive grove. Issued with British Military Authority money at a fantastically beneficial exchange rate, we were encouraged to fraternise with our erstwhile enemies who had not as yet formally surrendered but who welcomed us all like long lost friends. With Asti Spumante at a few pence per bottle sobriety was not the order of the day, particularly when surrounded in local bars by scores of Italian soldiers and sailors all seemingly anxious to demonstrate their undying friendship for the British Army who had freed them from the wicked Germans! Little needs to be said about our sojourn in southern Italy. Although I have at times described it as 'Our Italian holiday' that was far from true for other elements of the 1st Airborne Division who made contact with the German forces and in so doing lost Major General Hopkinson, killed by a sniper. The relatively inexperienced Airborne Forces could ill-afford such a loss.

We had meantime moved up-country in the vicinity of two small towns, Noci and Putignano where 1st Battalion took-over the Villa Monte as its HQ and here I remained until November, still 1943, when orders were received for the Division to return to the UK. Hurrah, Home for Christmas became the catchwords of the day but, alas, for many of us this hope was not to be fulfilled. The Unit embarked on two ships at Taranto and duly set sail for home. Our particular ship, an ex-banana boat, developed engine trouble and despite berthing in Sicily for repairs we got no further than Bizerta where we disembarked and entrained on a narrow gauge, single track railway for Blida, south of Algiers. The distance was about 450 to 500 miles which took an incredible three days of cramped travel in cattle trucks, in which it was necessary to sit up to sleep, toilet facilities were non-existent and such needs were met by impromptu stops followed by mass squatting alongside the railway line. When the driver blew the train whistle it was imperative to rejoin the train immediately irrespective of any other considerations! At Bone, a principal station en-route, a group of soldiers marooned in the sidings in similar trucks and being fed by 'the locals', pleaded with our engine driver for their trucks to be attached to our train. Deemed to be impossible due to a possible overloading of our own 'pint size' engine, they waved us a mournful good-bye. I sometimes wonder if they are still there!

We set-up camp at Blida under canvas and for those of us who had experienced the misery of Tizi those many months ago, there was a distinct feeling of 'deja vu', compounded by the realisation that Christmas was fast approaching and we were far from home. The winter rains, torrential at times, did nothing to improve morale and had we known that our fellow glider pilots on the 'other' ship were docking at Liverpool on 10th December followed by Christmas leave there might well have been a riot. Two subsequent incidents illustrate that this is no idle supposition. We embarked for home at Algiers on Christmas Day standing in pouring rain and some three or four inches of water on the quayside watching officers' batmen carry seemingly endless baggage aboard, some in wickerwork hampers labelled as medical supplies. Glider Pilots were certainly not stupid and it was evident that most of this was 'personal loot', probably legitimately acquired but its priority loading a thoughtless provocation. The officer supervising this exercise slipped and fell into the sea betwixt ship and quay. A cheer went up; not one of us stepped forward to rescue the unfortunate individual who was eventually pulled to safety by fellow-officers. Belatedly we boarded the vessel, proceeded to our mess-deck, wet, cold and hungry and were then ordered to fall-in with our weapons and proceed to the ship's armoury where they were to be stored for the duration of the voyage. As one man we all refused to budge one further inch. Threats and pleadings from officers were ignored till, finally, the ship's Captain came in person, had the good sense to appreciate the situation and gave us all a respite until the morrow. A wise man. Next day we happily disposed of our weapons as instructed and good order was restored. No wonder that Michael Briody, our much respected RSM, remarked to me many years later when he had retired as a Major, that the glider pilots were the finest bunch of men he had ever met and added, not well served by many of their officers. Incidentally had he been with us at that time (he was on the 'other' boat) such a situation would never have developed.

CHAPTER 9

The Elusive Bridge

Back in England after disembarkation leave, I was off to North Luffenham for a much needed refresher course on Horsa's and thence to a wartime aerodrome at Blakehill Farm, Cricklade near Swindon, where I resumed my normal duties as Staff Sergeant (Intelligence). At the time of our absence in North Africa and Italy, the Glider Pilot Regiment had been the subject of a reorganisation, one feature of which was to re-name the Battalions as Wings, presumably in an attempt to create a feeling of greater affinity with the RAF. Unfortunately and misguidedly, in the 'transformation' 1st Battalion became No 2 Wing and the 2nd Battalion, No 1 Wing, thus by inference becoming the 'senior' Unit. Did nobody in the hierarchy pause to consider the resentment that this would cause with the home-coming Battalion ? It certainly rankled with all of us and created an unfortunate feeling of having been 'stabbed in the back' by a certain senior officer generally assumed to be a close 'buddy' of Chatterton. I love the story told by Michael Briody which arose out of this illogical situation. The whole Regiment was to be inspected by His Majesty King George VI at Tidworth, and as the senior RSM, Michael Briody was in charge of the parade. Lt Col Murray, officer commanding No 1 Wing, had fallen his men in at the front of the parade. "Not so" said Michael. "No 2 Wing Senior Battalion Sir, pride of place, front of parade". In accordance with Army protocol as Parade RSM his word was law. The necessary re-deployment was effected!

One other organisational change was in the offing stemming in part from a general shortfall of trained glider pilots and in part from a recommendation of the late General Hopkinson following his experience as a second pilot with our C.O. – John Place – whilst in North Africa. It had been decided – wisely methinks – that a specific category of 'Second Pilot' should be created, necessitating considerably reduced training hours, culminating in the award of distinctive 2nd Pilot Wings qualifying the recipients for 'on tow' duties – not take-off or landing – and leaving their further training to 1st Pilots at Squadron level. The system worked well. The 'dilution' of 1st Pilots enabled a rapid escalation in the number of operational gliders available for operations – D-Day was not far off – and the majority of our '2nds', all of whom in my experience were 'as keen as mustard' went on to qualify for their 'full' Wings in due course. Johnny Gittings, a young lad under twenty years of age, became my co-pilot, keen, cool-headed and in due course a highly competent pilot. Tragically he joined the many others posted as 'missing presumed killed' at Arnhem so never had the opportunity to fulfil his great potential. He had hardly started 'to live'.

Shortly after our return we acquired a new Padre, Captain Chignell, known affectionately as 'Chig'. What a wonderful man he was. He was universally liked and respected, as the following extract from our demob magazine – Age and Service' dated 17th January 1946, records, *"We think everyone will be sorry the Padre had to leave us. Men of the Cloth are not conspicuously popular these days, but everybody had a good word for Capt Chignell..."* In due course he became yet another friend for life, a wise counsellor and a wonderful companion whether bird-watching – both he and his wife Pauline were highly knowledgeable – supping the odd 'pint' or so, or listening to his beloved Elgar and Beethoven. These were of course joys yet to come, far from evident in early 1944 where I will again pick up my story.

D-Day planning was limited to our 'C' squadron flying the truly massive Hamilcar with an all-up weight of some 15 tons when loaded with a Vickers tank or a self-propelled gun. Towed by Halifax's they were based at Tarrant Rushton and there I journeyed shortly before D-Day with 'sealed' orders showing the approach routes for the gliderborne invasion of Normandy. So concerned was my CO with the Top Secret information to which I was party that I was not permitted to leave camp except on official trips, I exercised around the perimeter track at Blakehill Farm and as D-Day drew even nearer I was given my own armed guard!

Viewed from a distant fifty or so years later this may seem to reflect a degree of paranoia but in 1944 absolute secrecy was an essential ingredient for the success of the Second Front and such security precautions did not seem out of place. I can recall but little of the planning details as operation after operation was planned and cancelled following the success of the 'Invasion'. Such activities became as routine as any other office job but fortunately the induced boredom and frustration was eased by a glider flying programme focussed on training my newly acquired second-pilot, Johnny Gittings.

One particularly exciting episode which led to a 'Court of Enquiry' involved a forced landing over and under high tension power lines in the vicinity of Cumnor near Oxford. In all probability I had not kept a sufficiently close 'weather-eye' on John's flying and we found ourselves endangering the tug and

---

***Opposite page:** Handley-Page Halifax glider tug, 1944.*

I decided to cast-off and make an emergency landing. Had I not been flying a Horsa the outcome might well have been less happy – possibly fatal – but such were the excellent flying characteristics of this glider that a perfect landing was made in a ploughed field, to the consternation of the ploughman just completing an immaculate day's work. In addition to us two pilots I had my Intelligence Clerk – Private Randall – on board. With remarkable aplomb he asked me if this was normal procedure, then pushing his Airborne folding bicycle across the field, mounted and pedalled away to make the necessary phone-call. I was cleared of dangerous and irresponsible flying and my co-pilot certainly benefited from the experience.

A further much needed reorganisation was implemented, this time in the field of Intelligence planning. At Divisional level this resulted in the addition of a GSO1 (Air) to the Staff and at Wing level a Captain (Operations officer) was added to our establishment. The officer concerned was designated as a 'Technical Adjutant', thus investing him with the full authority of the Commanding Officer. Squadron Intelligence officers were in turn added to their establishments and were functionally responsible to the Wing Operations officer. An Intelligence officer at Wing level, to whom I reported, held the rank of Lieutenant. Sadly as far as I was concerned nothing seemed to improve. Were these appointments seen as 'jobs for the boys' ? I fear so. Finally, enough was enough when a highly incompetent Ops officer compromised our Section security in such a serious manner that I requested an interview with the CO and asked to be relieved of my responsibilities and returned to Squadron. I was of course the 'founder-member' of the

---

Opposite page: *Albermarle tugs and Horsa gliders ready for a mass take-off, 1944.*

Section, had seen countless officers come and go, few of whom had made any solid contribution to our duties and, unbeknown to me, I enjoyed total support from the CO – John Place – and his 2 i/c – John Blatch. My request was 'not granted' but the officer concerned was speedily 'returned to Squadron'. At long last a dedicated and committed Intelligence officer was appointed, Lt Joe Maltby who also became the CO's second-pilot. He was an ex-Artillery officer with considerable experience in the field of enemy flak, having flown on many missions as a rear gunner and flak observer. He provided real leadership and I like to think that together we were an effective team. At last one of the many post-invasion glider operations moved from the planning stage to actual realization – or so it seemed – but many of the details have escaped my memory. Maybe I sub-consciously wiped my memory clean!

However the practical aspects are as vivid as on that evening in late June 1944. A small glider force was involved including many if not all of the HQ staff. Certainly the CO was due to fly one of the Horsas. My own glider was fully loaded with jeep and trailer containing a mixture of high explosives – aminol and plastics mixed – enough as John Place said to blow the whole glider formation out of the air should I be hit by flak. My complement included two RE officers and two RE other ranks, experts in demolition. The objective was the blowing-up of a bridge, or maybe two, over which it was feared that German armour would cross and thus threaten the post-invasion bridgehead. It was decided that I should take-off well in advance of the main party, late at night to enable a landing to be made at dawn. My glider, minus the two RE officers who had transferred to another aircraft, was towed to the head of the runway at Blakehill Farm, tow rope connected, Dakota revving-up prior to take-off when a red Very light was fired. Take-off abandoned – much cursing from both myself and John, 'fired-up' as we were for this adventure. Mercifully for us, photographic reconnaissance Spitfires from Benson, Oxfordshire had recorded the presence of German armour already in position on our designated LZ. A near miss!

And now to the tragic finale of this period of my life, the Battle of Arnhem – 'A Bridge Too Far' as it is popularly referred to. Early in September I was called to Moor Park together with the Intelligence and Operation officers, where planning was in an advanced stage for Operation Market Garden as it was to be called. My immediate reaction was that the proposed LZs were too distant from the single and essential objective, the bridge over the Lower Rhine at Arnhem. One of many reasons given for the location of the LZs was the supposed unsuitability of the terrain near the bridge, though the concentration of flak in that area was yet another consideration. For the second occasion in a planning situation I was disadvantaged by my lowly rank, but I pressed my misgivings to the point of asking who had judged the suitability or otherwise of the polders near the bridge as a potential LZ for at least a Coup de Main type of glider force. Apparently it was members of the Dutch Resistance, who clearly lacked the necessary experience and know-how to make such a judgement. I suggested that only a trained glider pilot could assess the terrain and volunteered to be parachuted in to make the necessary appraisal, but it was correctly adjudged that this was impracticable on grounds of timing, communication and security. Hindsight reveals that I was not alone in my misgivings. The ensuing battle and bitter defeat having filled many books and created two memorable films, I need only to place on record a 'worm's-eye-view' of the subsequent events

Johnny Gittings, the author's stalwart 'Second Pilot', only 19 years old, reported 'missing presumed killed at Arnhem, 1944.

as they affected 2 Wing HQ, Glider Pilot Regiment from Sunday 17th to Friday 28th September 1944. As the tragedy unfolded, these events were recorded in our Unit War Diary, a terse and official document which I was required to update day by day. I have used this source to prompt my failing memory as I write a somewhat more personal account of this distressing yet formative experience.

From March until August 1944, 2 Wing HQ together with 'F' Squadron had been based at RAF Blakehill Farm near Cricklade, Wiltshire. Here we trained with Dakotas of 46 Group whose primary function was to fly seriously wounded personnel from Normandy – post D-Day – back to a hospital which formed part of the aerodrome complex. By early August, in order to free much needed airspace at Blakehill Farm, we were moved to Broadwell near Brize Norton, Oxfordshire and it was from here that we emplaned for Arnhem, Padre Chignell 'thumbing' a lift with me at the last moment.

Throughout the previous day, Saturday 16th September, my second pilot Johnny Gittings and I had been fully engaged in supervising the loading of our Horsa Glider which was to carry a Border Regiment mortar platoon into the battle. By a strange coincidence some of my passengers had flown with me to Sicily in 1943 and expressed the hope that we would not repeat the previous experience of 'ditching' in the sea. I reassured them that such a prospect was certainly not part of the flight plan. As first pilot it was my responsibility to ensure that my load was safely secured and within the weight limit permitted, some 3½ tons giving an all-up weight of about 7¼ tons. In the event, additional equipment and ammunition was 'sneaked' on board resulting in a slight overload, a not uncommon occurrence.

At last after so many cancellations and postponements, the day dawned which was to see the launching of the combined American and British Airborne Forces in the greatest airborne operation ever conceived. The weather was not ideal. There was practically no wind and thick cloud extended from about 800 to 2,000 ft. Nevertheless Johnny and I, despite the weather and my own misgivings of the Battle Plan, kept our fingers crossed as we waited expectantly on the runway for actual take-off, that no red Very light would be fired on this occasion. We were away. The take-off was good and all 24 gliders from this 'drome got off the ground safely. Considerable difficulty was experienced in the clouds, many cable angle indicators – blind flying instruments – appeared to be unserviceable, as a result of which six gliders found it necessary to release and force land. These eventually followed on with the second lift which took-off on the following day. The cloud mass was rising rapidly, at times faster than we were climbing, so we kept popping in and out of masses of cumulus while circling the 'drome prior to setting course. Having passed the East Coast, clear sky was encountered, and the glider/tug combinations were able to close up into the intended formations. Meanwhile, Allied fighters were roaring overhead, circling and then going ahead to deal with any enemy opposition that may have been waiting for us. Judging from the fact that no enemy fighters were encountered, they must have fulfilled their mission and were also instrumental in silencing the majority of the 'flak' sites which otherwise would have given us a hot reception. As a result of this, 'flak' was slight, though some gliders caught rather a packet. Mine was one of these, inducing the Padre to fall on his knees in the cockpit and pray to the good Lord for deliverance. I told him that it was I, and not the 'good Lord' who would save him

– a story which he oft repeated throughout the many years of our ensuing friendship. Sadly our Commanding Officer, Lt Col John Place, who was flying the leading glider, was badly hit and his second pilot Lt Joe Maltby died of wounds as he flew on to Arnhem. Thus at this early stage I once again – shades of Sicily – found myself without an Intelligence Officer, a situation which was to have a significant effect both on myself and my Commanding Officer.

The Landing Zone proved to be excellent and we landed within a few yards of our objective. My passengers cleared the glider without incident and only desultory fire was encountered from the ground. Glider pilots moved off in their respective Squadrons and together with elements of the 1st Airlanding Brigade took up defensive positions around the Landing Zones which had to be held free of the enemy in readiness for the 2nd lift arriving on the following day. Meanwhile 1st Para Brigade were, unbeknown to us, having a tough fight to capture and hold the vital bridge at Arnhem, some 6 to 7 miles away to the East.

After a fairly quiet night the day dawned – Monday 18th – with every prospect of good weather. We were all expectantly awaiting the arrival of the 2nd lift so at 10 o'clock when the sound of fighters was heard, we all tried to get a good view of 'our boys' coming in. To our amazement Mersserschmitt 109s and Focke Wulf 190s roared overhead strafing as they went. After flying backwards and forwards over our area for about 20 minutes they left us alone and we began to wonder what had happened to the 2nd lift. By the afternoon, pressure had been brought to bear on three sides of our pocket and in one place we were forced to make a slight withdrawal. It was imperative that the area of the LZs was kept free of the enemy, so no move

On the ground safely, Horsa gliders after touching down in their Arnhem Landing Zones (above) and a Hamilcar with self-propelled gun emerging (below).

could be made to assist 1st Para Brigade who were having a tough time in Arnhem where the Divisional Commander had become temporarily missing. At about 1500 hrs, much to our relief, the 2nd lift arrived, having been delayed by bad weather in England. We then moved towards Arnhem. took up our position in a house and spent a disturbed night.

Tuesday, 19th September proved to be the fateful day of the whole operation. At 1100 hrs an attack was launched with the objective of entering Arnhem and joining forces with 1st Para Bde., who were desperately holding on to their occupation of the bridge. It proved to be impossible to break through a powerful wedge which the Germans had deployed on the western outskirts of Arnhem and thereafter we were pinned down in the suburb of Oosterbeek in a grim defensive battle which was to last for almost a week. Re-supply from the air was only partially effective due to intense 'flak' and severe RAF losses, but also by the lack of Dropping Zones (DZ's) due to our ever contracting perimeter.

It was on this day, Wednesday, 20th September, that the Germans began a full attack against the Oosterbeek perimeter. 2 Wing HQ had been forced to abandon their position in a house which by now was too far forward to be held and we withdraw southwards to a housing estate just north of the Arnhem to Renkum main road where we remained until the final withdrawal. Meanwhile the casualty rate amongst 2 Wing HQ officers was devastating.

In addition to the Intelligence Officer lost en route, our second-in-command had failed to arrive on the 2nd lift, his glider having been severely damaged by flak. He had forced landed and eventually found his way to Nijmegen. Captain Shuttleworth, our Adjutant, was killed by a sniper and his

Assistant Adjutant by shell fire. Both the Commanding Officer and the Regimental Sergeant Major were battle casualties and evacuated to a Field Dressing Station leaving me, a Staff-Sergeant as the senior remaining rank. Thus it fell to my lot to attend the Brigadier's 'O' Group on a daily basis. At these conferences, information concerning the state of the battle was exchanged and Orders issued for such redeployment as was considered necessary. As a consequence of these duties I maintained the 2 Wing War Diary which, as I have already stated, forms the background to this more personal account.

Although distances were short within the perimeter, speed of movement was a plus. I had been helped by retrieving a motor cycle from a Despatch Rider who had been killed by a sniper but alas I was soon to lose this mode of transport. Approaching Brigade HQ heavy mortar shelling was experienced and a nearby ditch was the safest place. I was within 50 yards or so of Brigade. On emerging from the ditch I found my motor cycle ablaze and on reaching the 'O' Group room found a charnel house with many wounded and four staff officers killed. Fortunately the Brigadier had escaped unscathed. Remarkably, some 43 years later I discovered that one of the four officers killed was the brother of our 2 i/c, Major Blatch, now one of my closest friends. By this stage of the battle we were undergoing intense shelling and attacks by both Tiger and Panther Tanks and armoured flame-throwers. Under such pressures we were continually having to shorten our defensive perimeter.

At 0330 hours on Thursday, 21st September, we again contracted the Divisional perimeter. The pocket was by now some 2,000 yards deep and 800 to 1,200 yards wide. The Germans used this to advantage. Mortars, rocket projectors and

2 Wing, Glider Pilot Regiment HQ, Arnhem, 1944.

artillery of all kinds simply pumped explosives into this area day and night for the next four days. 15 Tiger tanks were reported in one sector, armoured self-propelled guns were brought into action against us and German infantry tried infiltrating through our positions but were repelled. Reports were received that the 2nd Army was held up south of Elst and we began to realise the gravity of our situation.

Friday, the twenty-second was the day of the sombre message from the Brigadier. "We are up against it, but have to hold on at all costs." This was easier said than done when continual bombardment was rapidly reducing the number of effectives and no reinforcements were available. Nevertheless another German infantry attack was beaten off and morale rose accordingly.

We were losing count of the days as continual shelling took an ever increasing toll of our numbers. Food was non-existent and only rainwater from the roof had kept us going for the past two days but even this was all but exhausted. It was however, Saturday, 23rd September, the end of our first week engaged in this tragic operation. It was on this day on my return from an 'O' Group conference that I was confronted by a Divisional officer, threatened with a Colt 45 revolver and ordered to take command of a gun crew who were in a state of total panic. I saw no future in arguing, so escorted the crew back to their gun, learned that their sergeant had 'disappeared', appointed a corporal in his place and, with some stealth, departed on my way to transmit Orders for yet another shortening of our now overstretched line of defence.

It seemed an eternity since the previous Sunday, yet it was only 24th September. Shelling was intense. In one hour over 500 rounds were pumped into an area of barely 200 by 400 yards. We were forced out of the adjoining woods. The 40 or so houses forming our northern defence line had been reduced to no more than 30 by systematic burning and destruction and, to add to our discomfort, an attack supported by armoured flame-throwers had enabled Jerry to cover our line of retreat with machine gun fire. Between us and the main road, and therefore the rest of the pocket, was a Common some 200 to 250 yards wide. This formed a complete barrier to any further movement in daytime and the garden of our house was under fire from snipers. We were definitely pinned down inside our houses and holding on by the skin of our teeth.

Monday, 25th September. News came through that a Battalion of the Dorsets had made the crossing of the Rhine and was holding a bridgehead for our evacuation. They were none too soon. Not one house was intact and self-propelled guns were firing their 88mm shells through house after house with complete impunity. At 2100 hours, 2 Wing HQ commenced withdrawing and I had the unhappy task of telling our cellar full of wounded that they would be left behind to become prisoners-of-war. As I gave one last look around our house before sprinting across the Common, enemy infantry were already entering the garden. The Common proved a death trap for many, as 'Jerry' poured everything that he had into it as if he realised our plans for evacuation. My second-pilot, Johnny Gittings, who, young as he was, had proved to be a tower of strength throughout the battle, survived this final obstacle and was checked in by me at our evacuation assembly point. I never saw him again, no body has ever been discovered and I can but assume that he drowned during our crossing of the Rhine. All that remains of him is his name engraved as 'missing presumed killed' on a memorial at Groesbeek. So young...

On that same night I was slightly wounded and evacuated in a collapsible canvas boat, which was then sunk in the middle of the Rhine. It may be recalled that I had learned to swim after the Sicilian debacle, so I trod water and, remembering a First World War story when a British spy had saved his life by escaping into Allied lines by mouthing all the foulest obscenities that he could lay tongue to, I emulated his example. What a strange quirk of memory! Two Canadians were so impressed by this 'beacon' of invective that they put forth in a tiny craft, located me and hauled me aboard and landed me on the far bank. From here I crawled to a nearby Field Dressing Station and thus to safety. How fortunate I was.

Of the 10,000 original strength of the 1st Airborne Division, 2,000 were brought to safety that night. Of the 1,200 or so Glider Pilots who flew into Arnhem, 229 were killed and 469 were wounded or made prisoners-of-war. It was indeed a death blow for the Regiment in its original form. Thereafter it was restored to strength by the recruitment of volunteers from RAF pilots and this enabled Operation Varsity – the crossing of the Rhine – to be carried out in March 1945.

And so to the end of this sad and disastrous episode. On the 26th September we assembled in Nijmegen. The following day we were transported to Louvain in Belgium and on the 28th flew back to England, landing first at Bradwell and thence on to Membury and by truck to Broadwell. Our return is still deeply etched in my memory. The Mess staff had prepared a sumptuous feast for the returning pilots. Of the 120 pilots who had set forth some 12 days previously, just over 30 sat down for that meal. The empty places spoke for themselves. Many of the WAAF were in tears and despite our hunger, our appetites deserted us.

Much needed leave was granted to all concerned and I was joined by my ATS wife who had endured agonies of worry and suspense when listening to Stanley Maxted's vivid reports on the BBC. The aftermath of this disastrous defeat was yet another milestone in my Army life. It also, in due course, brought about a complete rethink of my post-war career.

As a direct consequence of the heavy losses sustained in the battle, particularly by 2 Wing HQ officers, I was encouraged and pressurised to apply for a Commission with the promise that only a six weeks course at the Officer Cadet Training Unit at Dunbar would have to be endured, due to my long service as a senior NCO. I should have long since learnt my lesson as regards Army promises – Despatch-riding and flying Army Transports should have sprung to mind. However, to a WOSB (War Office Selection Board) I duly went as a 'lamb to the slaughter', but that is yet another chapter in this saga.

CHAPTER 10

Some Thermal

So a month or so after the traumatic experience of Arnhem I succumbed to pressures from senior officers aided and abetted by my own 'delusions of grandeur', and set forth for a WOSB at Redhill in Surrey, secure in the assurances which I had received that should- ?- I be successful, the 'horrors' of pre-OCTU at Wrotham, Kent would not come my way. There was at that time a short 6 week OCTU at Dunbar for the likes of myself who had served as a senior NCO for the past 2 years. That was just about acceptable as far as I was concerned. It should be borne in mind that the pay of a S/Sgt Pilot was well above that of an Infantry 2nd Lieutenant with far less commitments in the way of Mess fees! Thus, an early return to the Glider Pilot Regiment as a Commissioned officer was an essential feature of my 'game plan'.

On the day preceding my departure to Redhill I returned to my quarters only to find the RSM, the Orderly Room Warrant Officer and my second-pilot, Sgt Lester, all busily engaged in smartening-up my wearing apparel – stripes and webbing gaiters blancoed, brasses gleaming, boots shining and other items immaculate in every detail. Never in my Army life had I reached such heights of perfection! Mike Briody, the RSM, who had often despaired of my somewhat 'scruff' appearance, declared that I was to do credit to the GPR and if I failed it would not be for lack of 'spit and polish'. Is it surprising that these friends and comrades became friends for life ?

WOSB (War Office Selection Board) which lasted for three days was a test of both social acceptability and leadership abilities. I was never entirely convinced which took precedence. The applicants, or should I say, supplicants were divided into groups, in our case five in number and a Military Testing Officer (MTO) was assigned to each group – in our case Captain Lord Walpole – to oversee our general behaviour even to the extent of accompanying us at times to the 'local'. Whether or not he bought his 'round 'I can no longer recall, but he was certainly an affable companion and totally free of any snobbishness as befitted his aristocratic standing. 'Dame Fortune' certainly favoured me in the qualities of my four companions in the three days of shared adversity. One was a paratrooper and an expert gymnast, two were members of the RASC Motorboat Unit – adept at the agility that this demanded and the other a tough ex-Commando with all the qualities you would expect from such an individual. At the opening session, not only were we introduced to the Major i/c the Board but also to several visiting Staff Officers from the War Office – impressive with their display of 'staff red-tabs' – who all appeared to our young eyes – I was 24 – - to be somewhat of an ancient vintage.

Initiative exercises 'in the field' alternated with written examinations, the former usually led by a nominated member of the squad save in the final exercise which was designed as a 'Group' effort. It was my misfortune to be nominated 'leader' for the first exercise which involved getting men and valuable radar equipment over a sunken lane, heavily mined by the enemy. The leader was given a fixed time to formulate his plans – something in the region of 3 – 5 minutes – which had to be limited to the use of basic materials at his disposal – for example ropes and scaffold poles. With little conviction I gave my orders and the crossing commenced. I overheard the resident Major confide in one of the visiting 'big-wigs' that the exercise had never been successfully completed using my option! I was furious. We succeeded and I felt a distinct feeling of 'one-upmanship' shared by the rest of the squad. An auspicious start to the days ahead. We had already endured a 'word association' assessment which consisted of flashing words onto a screen at rapid intervals – a few seconds – during which time the viewers had to write down their immediate reactions which owed nothing to logic but much to intuitive or sub-conscious reactions. The following day I asked (volunteered) to see the psychiatrist who I knew would be bound to be on the 'Establishment' at a WOSB – they always were. The particular individual was a Major who was obviously interested to see the strange individual who had had the temerity to volunteer to see him. His opening question was to ask me why. I explained that popular opinion in the Army portrayed the Psychiatrist as the 'bogey man' of WOSBs and I liked meeting 'bogey men'. He was highly amused, asked me to be seated and then referred to my word association test which had brought forth a spate of biblical verses and snippets from hymns. I explained my Plymouth Brethren background and this so fascinated him that the morning sped away, lunchtime loomed and we parted with a handshake. As I left he, laughingly suggested that the rest of my squad owed me a 'pint' as he, the 'bogey man' had no time left to see any of them. What a fascinating experience, demonstrating how deeply one's childhood training is embedded in the mind.

Other field exercises followed, led by the other members of our squad and, as far as I recall, all were executed satisfactorily. Come the last day and the 'group' exercise, to be followed by a 'special diabolical' test in front of all the resident staff plus the visiting dignitaries. The simulated scenario was that of a prison camp surrounded by a high wall – consisting of railway sleepers – some 12 feet high and topped by a high voltage single strand wire. We were walked through a hut full of diverse bits and pieces, including the inevitable scaffold poles, ropes of all types, broken ladders and the like. Anything from this store could be used for our escape but it had to be taken over the wall with us and hidden from the gaze of the sentries. Our squad, thanks to our gymnast, decided that one pole and a piece of rope would suffice and the subsequent 'escape' went without a hitch as if it had been rehearsed over and over again. Apparently we broke the WOSB record with our achievement – quite a 'feather in our collective caps'.

The last test now awaited us. All the squads now assembled in the hall of the fine old building that housed the WOSB, seated in the centre surrounded by the staff and visiting 'bigwigs'. The Major then announced that in turn we would all be asked to give a three minute impromptu speech on any subject of our choosing, and to view it as a privilege analogous to a 'Maiden' speech in the House of Commons! Bearing in mind the presence of Lord Walpole I ventured to intervene that surely it should be the House of Lords – much laughter. Inevitably as a 'W' my turn came last and I delivered a recruitment appeal for the Glider Pilot Regiment but at the same time pointing out that the assembled officers were too old for such activities. Once again loud laughter, my ordeal was over and I was free to return to Unit.

I awaited the outcome of my application with interest but no trepidation, as I was not that anxious to sever my long-standing associations with so many NCO friends. However, in due course the report arrived in our Orderly Room. I had passed but, horror of horrors, I was posted not to Dunbar but to the dreaded Wrotham. With the connivance of my good friend Joe Lowe – i/c Orderly Room – the posting was conveniently 'lost' and life carried on as normal until a few weeks later I was ordered to appear before our Brigadier – Chatterton – who was visiting Broadwell. He was in receipt of a critical letter from the Under Secretary of State for War, bemoaning the fact that Army resources were being wasted on the likes of me attending WOSB, being accepted for officer training and then failing to turn-up at OCTU. I was asked to explain this dereliction of duty. Knowing my Brigadier's total commitment to our Regiment I realised that I was 'on a winner' if I 'played my cards' right. I explained, quite sincerely, that after serving three years as a Glider Pilot I had no wish at this late stage in the war to be posted elsewhere, hence my decision not to proceed with my aspirations to become a commissioned officer. He was obviously sympathetic to my point of view and when I suggested that perhaps I could reply to the Under Secretary, after some pondering on this highly unorthodox request he agreed on the condition that the letter was passed through him as Commander Glider Pilots. Such a letter once composed, was typed by Joe Lowe and despatched as ordered. Weeks elapsed during which time we moved from RAF Broadwell to a 'non-flying' camp at Mushroom Farm near Braintree, Essex, where we were billeted in pre-war mushroom growing houses in stark contrast to the accommodation which we had left.

All thoughts of the letter had long since passed out of my mind when, one morning, an excited Joe Lowe dashed into the Intelligence Office to inform me that a letter had been received from the Under Secretary of State for War who had considered my letter and decreed that I was to be Commissioned immediately. Shortly before, this our CO, John Place, had been relieved of his Command and a Squadron Commander from No1 Wing, Major S.C.Griffith, had been appointed to succeed him with the rank of Lt Colonel. He had newly arrived and had had insufficient time to acquaint himself with the 2 Wing HQ staff. I was summoned to his presence, where I was informed in no uncertain terms that he disapproved of the recommendation of the Under-Secretary but that he had no alternative but to comply, I was Commissioned 'on the spot', supplied with the necessary clothing coupons for my new kit, given 10 days leave and informed that on my return I would be posted out of HQ as I had been there too long to function as an officer alongside my erstwhile NCO colleagues. With mixed feelings I duly departed on leave having contacted my ATS wife's HQ for her to join me. Off to London where I first visited Moss Bros and, to my good fortune, was kitted out with a pre-austerity service dress, made many years previously for a young officer of my precise build who was killed in action. Walking up Charing Cross road I was given a magnificent salute by a passing Guardsman prompting me to look around to see who he was saluting. Me of course -a strange experience. At Paddington Station I met a bewildered L/Cpl wife who failed to salute me but was excused this oversight. The date was the 27th February 1945 and she was by now some three months pregnant and would be demobbed in the very near future.

On returning to HQ I was speedily posted to an OTS (Officers Training School) where for three weeks I was to be inculcated with the mysteries of the Officers' Mess, the Loyal Toast, which way round to pass the Port etc., plus the rudiments of other necessary skills -firing a rifle and revolver, riding a motor cycle, basic parade ground drill and such things. Before I departed, my pre-austerity Service Dress was the envy of the Mess, not least the CO, who was most impressed by my sartorial splendour. The OTS was housed in a magnificent house, Rushden Hall near Kettering and was commanded by an urbane and affable Major, a professional Bridge player in civilian life who regaled us all with many a fascinating insight into pre-war high society. I was most fortunate. The majority of our Course consisted of middle-aged Honorary Captains all of whom were either ex-tea or rubber planters from Malaya. Their plantations were in the process of being recaptured from the Japanese and their return to Malaya was imminent in order to restore the infra-structure of the country. Their induction into the commissioned ranks of the Army was to enable them to integrate with the British Occupying Forces. The rest of the Course were Intelligence specialists and I was therefore the only combatant soldier amongst them. It was soon apparent that the 'field' aspects of the training were irrelevant as far as I was concerned and when on a map-reading exercise I led my group back to base long before anyone else put in an appearance – the issue of the 'master' map to me helped no end – I was invited to join the Staff for the rest of the time. Such an invitation could not be refused and for the remainder of my time I did my best to instruct the Honorary Captains in the arcane 'arts of war' and won the thanks and congratulations of the Major Commandant. I presume that a glowing report heralded my return to Wing HQ where, despite his previous misgivings, my CO – Billy Griffith of eventual MCC fame – retained me as Intelligence Officer. At this time I renewed my links with Peter Scott-Malden, the I.O. who had in 1943 been left behind at Matlock as we departed for North Africa. He was now Captain Operations officer, still my senior, a highly intelligent and efficient officer under whom it was a real pleasure to serve. Our relationship lasted but a few weeks. Early in June he was posted to our Indian Squadron and almost immediately and to my genuine astonishment, Billy Griffith promoted me to Captain to fill the vacancy and in due course I was confirmed as Operations officer. Thirteen weeks had elapsed since I was a Staff Sergeant – what a rapid uplift, some 'thermal'. Throughout all of this time my previous NCO friends were absolutely fantastic in their support for me in my new role. I can well remember the RSM, Mike Briody, passing me in camp, 'throwing up' a magnificent guards' style salute and accompanying it with a surreptitious wink that 'said it all'.

During my absence on leave and at Rushden Hall, 2 Wing HQ and 'F' Squadron had exchanged the discomforts of Mushroom Farm for RAF hospitality at Keevil, an aerodrome near Trowbridge, Wilts where I renewed my association with the Waco glider – now renamed the Hadrian – which I had last flown in North Africa. I was amused but also delighted that the Course Commander, an RAF Squadron Leader, deemed me to be 'proficient as a Hadrian pilot'. I doubt whether he had ever been faced with landing one in a choppy sea at dead of night!

My exalted rank now bore fruit in a very tangible manner. For the first and last time in my Army service I secured that much cherished benefit, a 'Sleeping-out pass'. My wife had been demobbed – very pregnant – and she joined me for two or three weeks in a delightful cottage in the nearby village of Steeple Ashton, inhabited by an equally delightful 'old' lady who fussed over Hilda as if she was her own daughter. All good things come to an end and this happy interlude was terminated by the end of the Course and our return to RAF Blakehill Farm on the 15th June 1945. It was almost a year since I had been stationed there, a year of collective tragedy for the Unit but a period in which I had made some modest personal progress.

I was by now formally confirmed as Wing Operations Officer with the standing of Technical Adjutant, a promotion much resented – and understandably so – by many officers who were many years my senior by virtue of their length of service. In one instance this manifested itself in blatant insubordination which I dare not ignore. Billy Griffith gave me his unqualified support, the offending Captain was posted and reverted to Lieutenant. I had no further problems of this nature.

The war in Europe was over, VE Day had been celebrated but Japan had yet to be defeated. Rumour, never very reliable, suggested that some of us were to be posted to the Far East and when on 3rd July I was detached to RAF Ibsley in order to learn the 'art' of Glider Snatch, the possibility of flying in and out of jungle clearings in Burma loomed large on the horizon. After the initial shock take-off, contrived by my instructor S/Sgt Westerby, a friend from Burnaston days, all went smoothly and in ten days I was duly 'certified' as a competent Snatch pilot for both light and heavy Hadrians. 'Snatching' involved a specially equipped Dakota flying in low at about 150mph, trailing a hook attached to a steel cable, engaging this with a loop and nylon rope attached to the glider and thereby 'snatching it off the ground in about 70 yards. It was exciting and, thanks to much experimentation by 'test' pilots, relatively safe. In fact I used the technique to retrieve a Hadrian which had forced-landed in the village of Minety near Malmesbury. It was a fairly large field hemmed in with mature beech trees and this required a carefully judged and 'flat-out' snatch, testing the skills of the Dakota to the limit. It was successful and, from the ground, spectacular, so much so that village gossip insisted that one of them new fangled jets had taken off from farmer so and so's field! On the 14th August Japan surrendered, although VJ Day was not celebrated until 2nd September. It was still 1945. Obviously my acquired skills as a snatch pilot were not going to be required in the Burma jungle and I shed no tears on that account, but an uncertain future lay ahead. The anniversary of Arnhem all but coincided with the birth of our first-born, a son, Martin, who the Padre insisted on calling Arnhem-Oosterbeek even when he met him in later years. Hilda was seriously ill in labour and the CO insisted on supplying me with his driver and a car for a visit to her bedside in Stourport-on-Severn, where, temporarily, she was living with her parents. Yet another milestone in my life. It was no surprise when early in November we left Blakehill Farm for the second and last time bound for an ex-RAF aerodrome, Finmere near Buckingham, where glider pilots from all Squadrons were being assembled, some for demob and others for an eventual posting to Palestine where they were to assist in the control and policing of Jewish immigration.

Our 2 Wing Headquarters staff were a friendly bunch of people, formality was kept to a minimum and cricket dominated many Mess conversations. The C.O., Billy Griffith was destined to become Captain of Sussex CC to be followed by his appointment as Secretary of the MCC. His second-in-command, (Major) Hugh Bartlett was a brilliant left-handed batsman and had won the Lawrence award for the fastest century in the 1939 season. Padre Chignell was the historian of Worcestershire County Cricket and in due course became

President of their County Club. In such august cricketing company I was an also-ran, never having played the game, except on the beach, in my life. Nevertheless I had been 'appointed' Sports' Officer, an entirely nominal, unpaid, dogsbody job by Billy who obviously thought it was high time I did something really useful. I did in fact represent the Unit in cross-country running and had the reputation of being super-fit!

It was all too apparent that 'Intelligence' activities were to have no relevance to the Finmere situation with war as a distant memory. Despite my protestations, the C.O. appointed me Education Officer with responsibilities for organising an Educational and Vocational training Establishment on the lines of an Army Formation College. Books, stationery and equipment were to be made available from Army sources but lecturers and instructors had to be 'home-grown'. Fortunately, by its very nature, our Unit had no shortage of talented individuals, and a wide range of classes was soon organised. 'Age and Service' was created as a weekly newspaper – the forerunner of 'Eagle', our current Association Magazine – and some extracts from an article headed, 'What – No School Today? ' by Capt.Waldron, Education Officer, in the 2nd issue dated 12th December 1945, set the scene.

"Three weeks ago, I introduced myself as the Wing Education Officer and promised to do everything in my power to satisfy the Wing's educational needs. This meant that I was faced with a dual problem. Firstly, a section capable of dealing with all the various demands of education had to be formed from the old Intelligence Section which consisted of one officer, one NCO and one clerk! Many appeals were made and as a result of the excellent response we now have a section of four officers, seven NCOs and three ORs who have been working strenuously and have produced among other things this paper you are now reading. Secondly, it was necessary to assess the Wing's needs, assemble a group of instructors, work out timetables and compile lists of text books and equipment. These last have been our bottleneck...Stationery for the school was drawn on Tuesday, books are being collected from the Command Library, and indents for equipment and materials for the Hobbies Section are being relentlessly 'chased'...The success of the actual school depends on your enthusiasm." In the event although it fell somewhat short of our over-optimistic aims, it did provide a much needed focus and stimulus throughout the mind-numbing period when men were impatiently waiting for their demob number to be reached. However, within six weeks of setting-up this project, Billy Griffith asked me if I would prefer a secondment to the RAF in preference to a posting to Palestine. My answer in favour of the RAF appointment was spontaneous and I have lived long enough to thank Bill many times over for his support in securing for me an appointment as GSO3 (Air) to 1333 TSTU (Transport Support Training Unit) where I assumed responsibility for all aspects of glider flying, glider maintenance and lecturing to senior RAF and Army officers on the use of gliders in WW2.

I have always assumed that my pre-appointment interview at Group Headquarters (46 I imagine) contained a somewhat unlikely story concerning an unacceptable level of fatal accidents at Syerston. Clearly, in peacetime these were difficult to explain away to grieving parents. Whatever the truth of this, suffice it to say that I was granted overriding powers to rectify the situation and that I was responsible directly to Group for the execution of my duties. It would have been helpful to say the least if my position had been clarified with the Group Captain

Commanding Syerston and his senior officers, in particular his Wing Commanders Flying and Maintenance. I was to replace a Flt Lt Wilson and to 'inherit' his supporting officers, Flt Lt Dave Marande and FO 'Andy' Andrews. The Warrant Officer in charge of glider maintenance was also to report directly to me. Come the day – 20th January 1946 – I arrived at Syerston, a peacetime aerodrome made famous by Guy Gibson and his Dam-Busters, was shown to my bedroom and introduced to my WAAF Batwomen and left to my own devices. No arrangements were made for me to meet Groupy or any of his senior officers and as it was lunchtime I made my way to the Officers' Mess, where I waited at the entrance to assess whether it was 'waitress or self-service'. As I stood there a Flying Officer wearing the half-wings of a Flt Engineer left his table, came across and enquired in the most friendly fashion whether he could help and in response to my question told me that it was a self-service Mess. I collected my meal under his watchful and helpful eye and we then shared a table throughout the meal. So I first met David Cripps yet another life-long friend our lives having 'intertwined' over many, many years, in business and in pleasure. For some years we were even next door neighbours!

After lunch David took me through the Station, pointed out the key buildings and led me to my own HQ located in a Nissen type hut. We parted company and I opened the door and met my staff-to-be for the first time. It was not an auspicious occasion. They were lounging around in – dare I say – typical RAF 'scruff', the coal-fired stove in the middle of the room had shed its ash outside the hearth and over the floor and the predominant decor was that of dirty NAAFI cups. Nobody moved and not for the first time in my Service life I was faced with a situation which I was not prepared to tolerate. What I saw epitomised the very slack organisation as described to me at my Group interview and I resolved to 'take the bull by the horns'. Did they realise that I was their newly appointed Commanding Officer ? They nodded their awareness of the fact as if it was of little consequence. I riposted swiftly and sharply, making it abundantly clear that I was not prepared to operate in a 'shit heap', that I would return in one hour precisely, there would be no ash on the floor, no dirty cups in sight and that they would be properly dressed, caps on and salute me as their CO when I next entered the room. I turned, and for an hour on a cold and miserable January day paced around the Station familiarising myself with the lay-out and wondering what the hell I would do if I returned to the same situation that I had just left. In short I had placed myself in a potential cul-de-sac. As it transpired all was well. The erstwhile filthy room was in near pristine condition, the occupants had buttoned-up their tunics and were wearing their peaked caps and rose as one man and saluted me, somewhat over-elaborately, as I entered. I thanked them, suggested we should all be seated asked for their first or nick-names and dispensed with all formalities. I had made my point and was not foolish enough to try to superimpose Army type discipline in an RAF situation. They proved to be a loyal and first-rate team who responded superbly to the changes which had to be effected if I were to fulfil the objectives set by Group and gave me full support in the turbulent days which lay ahead.

To return to that first afternoon and the serious business of investigating all aspects of the 'local' organisation. Flying Standing Orders, the Bible as it were covering all organisational aspects of actual flying, proved to be a woefully inadequate document, lax in the extreme not least in the all-important area of tow-rope dropping and retrieval. Over the next few days I rewrote these Orders from cover to cover. Of even greater importance as regards flying safety is the level of maintenance of the gliders, wooden aircraft requiring specially stringent standards. Each aircraft has its own maintenance log-book known as Form 700 and a reading of these revealed that certain key checks had not been carried out and the airworthiness of at least some of the Horsas – at this time all Mark 1s – was suspect. A visit to the maintenance hangar was next on my list and on arrival I was greeted with due decorum by a 'Chiefy' (Warrant Officer) wearing WW1 medals. He confirmed that 'wooden aircraft inspection standards' had not always been applied due to insufficient time being allocated for such work and he expressed his relief that at last someone in authority recognised this problem. I made arrangements to carry out airtests on three selected Horsas on the following day and as the local weather had prevented flying at that time I dismissed my staff and retired to my room to ponder my next move.

The airtest flights on the following morning confirmed my worst suspicions. Even as an experienced pilot I found the Horsas difficult to handle. In particular, aileron controls were too slack and unresponsive. Bearing in mind that 1333 TSTU was essentially a Training Aerodrome for both Tug and Glider pilots, the latter only having just emerged from their initial flying course, it was of paramount importance that their aircraft should be in 'first-rate nick'.

Without further ado, I endorsed all the many Form 700's as 'Unserviceable' and by means of the Station Tannoy announced all glider flying cancelled for three days. It was tantamount to 'lighting the blue touch-paper' and as I expected within minutes an irate Wing Commander (Flying) summoned me to his office where he sought to use his superior rank to override my 'outrageous' action. Faced with every Form 700 endorsed 'Unserviceable' I invited him to use his rank to counter my signature. This he declined to do but instead asked the Wing Commander i/c maintenance to join the discussion and to override my decision. This he also declined to do, so to resolve the impasse that my action had created, Group HQ was phoned. The outcome was as I had hopefully anticipated. The 'special' powers attached to my appointment were explained and a furious and chastened W/Cdr had to defer to my judgements. I remained at Syerston until 10th October, a stint of some ten months and in all of that time no human being suffered as much as a cut or bruise, one clash of wing tips on the ground caused only minor damage, we increased our flying hours both day and night by some 20% and the W/Cdr became in due course a staunch ally and even qualified as a Glider Pilot. A happy ending to the story after a stormy beginning. My 'popularity' was assisted as I had been allocated a 'personal' Miles Magister, an aircraft I had not flown since Burnaston days way back in 1942. The Squadron-Leader CFI (Chief Flying Instructor) gave me a refresher course and from then onwards, particularly on a summer's evening, I would take to the skies and indulge in my favourite activity, aerobatics. There was always a queue of tug pilots seeking to come up with me and thus relieve the boredom of their day by day towing duties. I rarely needed to buy a drink in the Mess!

A three days interlude in June, or possibly July, posed a very different challenge to that of flying. I was very close to my demob date and had made contact with the Admiralty Department that I had walked out from those many years ago.

They had moved from Bath to Harrow. Although I had been promoted to Executive rank it was a temporary war-time appointment and there was no certainty that come post-war cut-backs in the Armed Forces, I would not have had to revert to my previous position as Clerical Officer. This possibility was definitely not acceptable, but to ensure that I retained my Executive status it was necessary to sit and pass one of the Civil Service Post-war Reconstruction Examinations for the Executive Grade. I applied to enter the 2nd of such exams, where 1500 entrants competed for 60 vacancies – some contest. Formal study for this challenge was out of the question so I departed for Bedford, my designated exam centre, with little hope of success. I stayed at a local hotel, fortunately in sight of the exam venue, and on the preceding evening went 'out on the town' and met a group of Paratroop Officers with the inevitable consequences. I woke the following morning with the 'mother and father' of all headaches, partook of ersatz coffee and a little dry toast and made my way to 'sit' the key paper of the whole exam, 'General Knowledge,' a 3 hours test! 'The divinity that shapes our ends' once again came to my rescue. I was well read as regards the many magazines and journals of that era and an invitation to write on the significance of such publications was seized upon with glee. A question concerning the political 'hot potato' of the day, Nationalisation, was equally acceptable and other questions, of which there was a choice, gave me no problems. I recall that I wrote 21 pages of foolscap, just failed to conclude my last answer, came 2nd in the results with 96%, beaten by a girl entrant and went on to secure 32nd place overall and thus ensured the retention of my Executive position and was entitled to select another Ministry other than the Admiralty. The up and coming Ministry of Transport would have been my choice but to the consternation of my mother and sister, the amazement of my erstwhile Admiralty colleagues but the placid and active support of my wonderful wife, I decided to leave the Civil Service and take my chance in the strange and 'foreign' world of Industry, a decision which after some interim misgivings, I never regretted. That however, is a story not relevant to this particular saga.

Back at Syerston life continued in a fairly routine and organised fashion. My experience in shaping the day to day activities of this aerodrome had a direct bearing on the decision to which I have just alluded, to leave the Civil Service. The bureaucratic life, safe as it undoubtedly would have been – summed up by 'In tray', Out tray' and the all-important 'Pending tray' – was not for me after the challenging experiences of my Service life, and fortunately my perceptive wife shared my viewpoint. My demobilisation number 'fell due' in July but in response to a special request from the RAF I served an additional three months during which time efforts were made to persuade me to transfer to the RAF with the rank of Flight Lieutenant and a permanent peace-time Commission. I was sorely tempted but common-sense prevailed. Neither my wife or I would have been temperamentally suited to Service life in peacetime nor would my leftist political convictions been compatible with a successful Service career. So on 10th October I attended at the Regimental Depot, Fargo Camp, for the last time and by declining an 'invitation' from the Depot Commander to attend at a Mess Dinner with the GOC Southern Command, I immediately became 'persona non grata' and ensured that I departed my Service life in an aura of hostility. By borrowing an NCO's greatcoat I was able to spend the evening in nearby Salisbury with some of my old NCO friends and when, the following morning, I was refused transport to get me to Aldershot, by a bloody-minded Adjutant, S/Sgt Sharpe, a long-standing ex HQ member of staff, now i/c vehicle maintenance, took the CO's car for a road test with me on board. By pure chance (?) we finished up in Aldershot where I collected my various demob clothing and re-entered Civvy Street after a break of six interesting and at times, turbulent years. The final episode was yet to come, a 'sting in the tail' that contained all the necessary ingredients for a controversial conclusion to a controversial Army career. Shortly after confronting the stark realities of 'Civvy street', the letter from the Army Paymaster's Department that dropped through my letter box was a welcome sight, presumably containing my final pay cheque and Demob Bounty of £200 or so – a princely sum in 1946. To my horror and amazement it enclosed a curt and formal demand for a refund of monies overpaid to me over a period of three months. For the first time in my Service life I was in debt to the Army!

Needless to say written clarification was sought for this sorry state of affairs and after much evasion the 'truth was out'. It will be recalled that at the RAF's request I had served an additional three months beyond my official demob date and this provided the opportunity for bureaucratic skulduggery at the hands of certain HQ officers who had long resented my rapid elevation to Staff Officer. A back-dated Southern Command Order demoted me to Lieutenant – non-flying and non-Staff – from July to October, the additional three months, effectively reducing my pay by two-thirds! Small wonder I owed the Army money.

Luckily it so happened that my local M.P., a typical bumbling ex- Colonel, was visiting Stourport-on-Severn – my new home – for a meeting with constituents so to him I presented my problem. I played the 'card' of 'Service to King and Country' quite shamelessly and he responded by becoming suitably enraged at my shabby treatment. True to his word, at 'Question Time in the House', he put my case to the Secretary of State for War from whom I duly received a letter of apology, re-instatement of all the disputed monies and the grant of the title of Honorary Captain for life. No less a personage than the Minister himself came to my aid in this my final battle with the 'Establishment'.

My view from the cockpit of the towing aircraft as I took my last glider flight.

EPILOGUE

End Or Beginning ?

The remaining words can only be written when distance – in excess of 50 years – enables me to perceive that a jumble of seemingly unconnected events contained within it a formative journey from the uncertainties of adolescence to the confidence, however misplaced, of adulthood. I now see this development process as a three-stranded cord, intertwined to the extent that the three separate strands are barely discernible as such, each inter-dependent on the other. I have focussed my mental spotlight on the nine years from 1937, when I left home not yet 17 years of age, to 1946, when buttressed by the self-confidence engendered by experience, I forsook the safe haven of the Civil Service for the uncertainties of Industry.

The least discernible of the 'three strands' was my gradual, almost imperceptible, introduction to wider social and cultural influences which were denied me throughout my childhood within the 'Closed Order' of the Exclusive Brethren. From the moment I left my home I 'rubbed shoulders' with all manner of men and women with a kaleidoscope of backgrounds, beliefs and experiences. A delightful professional car thief, a bombastic Brigadier, the 'soon-to-be' Secretary of the MCC, a devout Padre, all contributed to a growing awareness of the richness and diversity of personal relationships with all manner of people. Hand-in-glove with these experiences came introductions to the world of literature, classical music, the theatre and concert hall and, on a few occasions to the world of prestigious London

Clubs! Such was the journey from the 'separateness of Apartheid'.

Inextricably inter-twined with this broadening of my personal relationships, came an awareness of the multi-faceted nature of beliefs- philosophical, political and religious. The latter inevitably led me to challenge the validity of my Christian Fundamentalist convictions. These, linked as they were to a dogmatic acceptance of 'Truth' as enshrined in every word of the Bible, both Old and New Testaments, left little room for compromise. My decision to volunteer for the Army, seen in this context, was thus a head-on clash with my dogmatic beliefs, not a matterof 'degree' but one of 'absolutes'. In such a conflict the Bible and its enshrined 'Truth' lost out and, without conscious thought, the very foundations of my 'faith' were eroded. My subsequent association with the Communist Party was a direct consequence of this thought 'vacuum' and although of brief duration – 1943 to 1944 – my introduction to Marx, Engels and a raft of other philosophers has had a profound and lasting influence throughout the rest of my life and has provided the philosophical 'under-pinning' for my continuing socialist convictions. At the same time, little by little, the dogmatism which had been an overwhelming influence throughout my childhood, was being supplanted by a much greater sense of tolerance – not the tolerance of condescension but that of awareness and sensitivity – deriving from my close friendships and association with individuals of many 'faiths', or no 'faiths' at all; of diverse political persuasions and of differing concepts of ethics and morality. It became increasingly apparent to me that individuals epitomising sincerity and integrity and possessed of that most elusive yet essential quality, innate 'goodness', were guided and inspired by precepts rooted in a bewildering array of religious and philosophical beliefs. 'Exclusivity' no longer formed any part of my personal 'credo'.

Lastly, these nine years were in many ways an extended 'Management Course', developing such innate abilities as I may have possessed in the inter-connected fields of leadership, management and the organisation of scarce resources. I could not have joined a more effective and demanding 'University', combining as it did innumerable challenges with the wisdom and guidance of the very special individuals who helped me on my way at crucial moments – the milestones in my life.

Today, I am not what I was in 1946 and in closing this section of my life I must not superimpose upon a young man of twenty-six, the convictions, beliefs and attitudes which are the products of another fifty years of continued development, yet to be committed to paper. However one underlying conviction, as expressed in an earlier chapter, has remained a 'constant' throughout all these intervening years.

"For me, the humility of uncertainty is an immeasurably finer attribute than the certainties of dogmatic belief."

Operation Market Garden, 1944: *a selection of World War II military equipment that would have formed part of the payload of gliders taking part in the Arnhem airborne landings.*

operational
Z

Aerial reconnaissance picture of some of the Arnhem landing zones, showing the massed glider landings. The Author's aircraft and those of some of his friends are circled, as is the location of the battlefield HQ of 2 Wing, Glider Pilot Regiment.

Airspeed Horsa gliders assembled in formation on the runway and ready for take-off, prior to Operation 'Market Garden'.

*Mass glider landings.
Normandy, June 6th, 1944*

Glider training landings.
RAF Brize Norton
21st April 1944.